OCR

D1481785

DISCARD

THE COMMON
FISHERIES POLICY

Origin, Evaluation and Future

Mike Holden

With update by David Garrod
BSc, PhD
Director of Fisheries Research in England and Wales 1989–94

A Buckland Foundation Book

Fishing News Books

Copyright © Trustees, The Buckland
Foundation 1994

Fishing News Books
A division of Blackwell Science Ltd
Editorial Offices:
Osney Mead, Oxford OX2 0EL
25 John Street, London WC1N 2BL
23 Ainslie Place, Edinburgh EH3 6AJ
238 Main Street, Cambridge,
    Massachusetts, 02142, USA
54 University Street, Carlton,
    Victoria 3053, Australia

Other Editorial Offices:
Arnette Blackwell SA
    224, Boulevard Saint Germain
    75007 Paris, France

Blackwell Wissenschafts-Verlag GmbH
    Kurfürstendamm 57
    10707 Berlin, Germany

    Zehetnergasse 6
    A-1140 Wien
    Austria

First published 1994
Reissued in paperback 1996

Set by Florencetype Limited
Printed and bound in Great Britain by
Hartnolls Ltd., Bodmin, Cornwall

DISTRIBUTORS

Marston Book Services Ltd
PO Box 269
Abingdon
Oxon OX14 4YN
(*Orders:* Tel: 01865 206206
       Fax: 01865 721205
       Telex: 83355 MEDBOK G)

USA
Blackwell Science, Inc.
238 Main Street
Cambridge, MA 02142
(*Orders:* Tel: 800 215-1000
       617 876-7000
       Fax: 617 492-5263)

Canada
Copp Clark, Ltd
2775 Matheson Blvd East
Mississauga, Ontario
Canada, L4W 4P7
(*Orders:* Tel: 800 263-4374
       905 238-6074)

Australia
Blackwell Science Pty Ltd
54 University Street
Carlton, Victoria 3053
(*Orders:* Tel: 03 9347-0300
       Fax: 03 9349-3016)

A catalogue record for this title
is available from the British Library

ISBN 0-85238-242-1

Library of Congress
Cataloging-in-Publication Data
is available

Disclaimer

The opinions and interpretations expressed in this book are entirely those of
the author and under no circumstances can be interpreted to engage in any
way or necessarily reflect the position of the Commission of the European
Communities or any other institution of the European Communities. The
author is responsible for any errors of fact or interpretation.

A Pioneer of Fishery Research

Frank Buckland 1826–1880

The Man and His Work

Frank Buckland was an immensely popular mid-Victorian writer and lecturer on natural history, a distinguished public servant and a pioneer in the study of the problems of the commercial fisheries. He was born in 1826, the first child of William Buckland DD FRS, the first Professor of Geology in Oxford who was an eminent biologist himself. From infancy Frank was encouraged to study the world about him and he was accustomed to meeting the famous scientists who visited his father. Like many other biologists of his day, he trained as a surgeon; in 1854 he was gazetted Assistant Surgeon to the Second Life Guards, having completed his training at St George's Hospital, London. He began to write popular articles on natural history and these were issued in book form in 1857 as 'Curiosities of Natural History'. It was an immediate success and was to be followed at intervals by three more volumes; although long out of print these can be found in second-hand bookshops and still provide entertainment and interest.

His success increased demands upon him as a writer and lecturer and he resigned his Commission in 1863. He had become interested in fish culture, then regarded simply as the rearing of fish from the egg. This involved the fertilization of eggs stripped by hand from ripe fish with milt similarly obtained. Release of fry was seen as a means of improving fisheries, particularly of salmon and trout, in rivers and lakes which had suffered from over-exploitation or pollution. He gave a successful lecture on the subject at the Royal Institution in 1863, subsequently published as 'Fish Hatching', and was struck by the intense interest aroused by his demonstration.

He was permitted to set up a small fish hatchery at the South Kensington Museum, the forerunner of the Science Museum, and by 1865 had collected there a range of exhibits which were to form the nucleus of his Museum of

Economic Fish Culture. This aimed to inform the public about the fish and fisheries of the British Isles and for the rest of his life he laboured to develop this display. Although he was paid for his attendances at the Museum, the exhibits were provided by him at his own expense; in his will he gave the collection to the nation. Until 1866 he produced a steady stream of natural history articles mainly for 'The Field'. Then he helped to establish a rival journal, *Land and Water*, which he supported until his death.

National concern over many years at the decline of salmon fisheries, which suffered not only from overfishing and pollution but also extensive poaching and obstructions such as locks and weirs, led in 1861 to the passing of the Salmon Fisheries Act under which two Inspectors for England and Wales were appointed. When one of the original Inspectors resigned in 1867, Buckland was an obvious choice as successor. He had already accompanied the Inspectors on their visits to rivers and was also often asked for advice by riparian owners. He would think nothing of plunging into a river in winter to help net fish for the collection of eggs.

Britain's growing population in the last century created many problems of food supply; the sea fisheries offered a cheap source of abundant first class protein and as a result the marine fisheries, and particularly the North Sea fisheries, grew spectacularly. Little was known about sea fish; no statistics of fish landings were available, at least in England, and the biological basis of fisheries was a mystery, though it was widely believed that marine fisheries were inexhaustible. Nevertheless there were disturbing indications that previously prolific fisheries were no longer profitable and many Royal Commissions were set up. The most famous was that of 1863, which had Thomas Henry Huxley as one of its members. Buckland himself sat on four Commissions between 1875 and his death, a fact which reflected his increasing standing as a fisheries expert.

During his lifetime a number of public fisheries exhibitions were held abroad, and he tirelessly pressed for something similar to be staged in the United Kingdom. Unfortunately he died before he could see his wish fulfilled, but there is no doubt that the exhibitions held in Norwich (1881), Edinburgh (1882) and London (1883) owed much to the public interest he had worked so hard to engender. It should be noted that the Marine Biological Association of the UK, with its famous laboratory at Plymouth, was a direct result of the enthusiasm and concern created by the Great International Fisheries Exhibition held in London in 1883.

He died in December 1880, possibly of a disease caught from his parrot, for he had always been careless about his health and must have worked for long periods at full stretch to maintain such a high output of material. What he wrote was sometimes uneven but he was often breaking new ground. At all times he was concerned to explain, to teach and, most particularly to make the general public aware of the importance of their fisheries and the need to protect and develop this great national asset. A few days before his death he signed his will. His wife was to have a life interest in his estate but he bequeathed a sum of money which on her death should be used to establish a trust fund to support 'a professorship of Economic Fish Culture, to be called

The Buckland Professorship'. The main responsibility laid on The Buckland Professor was that lectures should be delivered each year at suitable venues in the United Kingdom.

The Foundation

It is clear that Frank Buckland intended the term Fish Culture to be widely interpreted and to cover much more than fish hatching and the rearing of fry. Consequently, when the original £5000 endowment became available in 1926, after the death of his widow in 1921, the original Trustees of the Buckland Foundation took a broad view of the subjects that Buckland Professors should be invited to write and lecture about. In 1930, for example, the first Buckland lectures were given by Prof. Garstang, a leading marine biologist of the time, on the subject of 'Frank Buckland's Life and Work'. The following year it was 'Salmon Hatching and Salmon Migrations' and after that 'The natural history of the herring in Scottish Waters'. As fisheries science and fishing methods evolved many subjects presented themselves which were unknown to Buckland and his contemporaries and succeeding Trustees have sought to ensure that such topics are covered so that the lectures have always been timely, important and of value and interest to those who depend for their livelihood on some aspect of fish and fishing. Each of the three Trustees holds office for a five-year period and the day-to-day business of the Foundation is managed by its Clerk.

In the Spring of each year there is a meeting of the Trustees at which the subject and the Buckland Professor for the following year are chosen. In accepting the invitation to hold office the nominee also accepts the responsibility for producing a text and giving at least three lectures at venues that provide the closest possible link with that area of fish and fishing being examined. The text has to be approved by the Trustees before the lectures are delivered. The Professor receives £700 for the delivery of the manuscript and £100, with expenses, for each lecture and a commemorative medal at the end of his year of office. More often than not the texts of the 43 Buckland Professors holding office so far have been published as books and copies of the more recent ones are available from Fishing News Books.

The Trustees feel that by continuing to keep alive, via the means willed to them through Frank Buckland's own inspiration, the memory of a man who dedicated his life to the improvement of the commercial fisheries of the British Isles they are helping, in their turn, to improve conditions in the present commercial fisheries. As the 50th Lecture in the series begins to come into view they are hoping for increased recognition of both Buckland, the Man, and of the Foundation he instituted.

Contents

Preface

Frank Buckland lived during a period when it was thought that the sea offered an inexhaustible supply of fish. But even towards the end of his life there were disturbing indications that this was not so. The plaice fisheries of the North Sea were becoming increasingly unprofitable: catches were falling and the size of fish caught becoming smaller and smaller. In response to pressure from the fishing industry, the UK government set up several Royal Commissions to examine the state of the industry and to determine the reasons for its impoverished state. Frank Buckland was a member of four of these Commissions. If Frank Buckland were alive today he would be shocked to find fisheries in no better state now than then. During his lifetime there was one good reason for mismanagement: ignorance of even the basic knowledge of the biology and life history of fish, let alone of what is known today as fisheries science. We now have a very good understanding of the biological principles which underlie the exploitation of fish stocks but we have been unable to put that knowledge successfully into effect, which is why Frank Buckland would be at a loss to comprehend why fisheries remain mismanaged. The economic management of fisheries, with which Frank Buckland was so much concerned, still eludes us. For this reason alone he would have been unlikely to have regarded the Common Fisheries Policy (CFP) favourably. The CFP, with which this book is concerned, is not a unique case; globally fisheries are mismanaged and there are few, if any success stories. What makes the history of the CFP so important is that it illustrates the extent to which the political background, against which decisions are taken, has prevented the CFP from being successful and, more importantly, looking to the future, how this background may prevent the policy from ever being successful.

One of the problems of writing about the CFP is that the term is frequently used to mean only the conservation policy, which is but one of the four, separate but interrelated policies which comprise the CFP. The other policies are those for structures, markets and international fisheries relations. There are many reasons for this, of which the most important is money. The policies for structures, markets and international fisheries relations provide vast amounts of money to the Community's fishing industry: in 1990 this amounted to ECU 429.41 million (£300.6 million) out of a total fisheries budget of ECU 446.31 million (£312.4 million), the balance being accounted for mainly by expenditure on control and monitoring. This is only Community funds; remembering that many measures stipulate that Member States must provide 50% of the funds, the real figure must approach twice this. Nothing is spent specifically on conservation measures, most importantly on compensating fishermen for the effects of their implementation. Every Member State has, therefore, an interest in supporting measures involving the policies for structures, markets and international agreements. Even if they do not gain from one of the policies, they gain from another and they will support those from which they do not benefit in order to ensure that they have the support of other Member States when the roles are reversed. For example, when the Commission makes a proposal for an agreement with Morocco concerning fishing rights in its waters, the agreement will be supported by Spain and Portugal whose fishermen are the main beneficiaries. Germany, the Netherlands and the UK may think that the price is too high but will not block the measure because they in turn will need support for structural or market measures. In contrast, all Member States will combine to block conservation measures because no Community finance is at stake. In fact, the reverse is true. Conservation measures inevitably concern limiting or even stopping fishermen's activities. They reduce fishermen's incomes. It is not surprising that most of the high drama and furious debate which is widely reported concerns the conservation policy which the brevity of newspaper headlines makes 'the CFP'.

There is also a particularly British reason why the conservation policy is considered synoymous with the CFP, although it applies to Ireland also. The structural and market policies were adopted before the UK acceded to the EEC and therefore the UK was not involved with the negotiations leading to their adoption. Even though the structural policy has played an important part in the development of the UK fleet and port handling facilities, the grants from the EEC are administered by the Sea Fish Industry Authority and, for this reason, many British fishermen think that they are made by the SFIA, not the EEC. The markets policy plays a very minor role in the lives of British fishermen because most fish are sold above the withdrawal prices laid down in Community legislation. The international relations policy also affects British fishing interests very little because by far the most important part of this policy concerns fishing agreements in waters off Africa where UK vessels do not fish. The policy also concerns agreements with Greenland, the Faroe Islands, Norway and Sweden but these are closely linked with the conservation policy because they concern total allowable catches (TACs) and technical conservation measures. For this reason they are identified closely by the British fishing industry with the conservation policy. In contrast, the UK was closely involved with the negotiation of the conservation policy and it was its attitude which was largely, although not entirely, responsible for the protracted negotiations. Furthermore, it has been the conservation policy, especially the system of TACs and quotas, which has made such an impact on the activities of British fishermen.

This is one of the reasons why this book is concerned mainly with the conservation policy although the other policies are not ignored. The second reason is that it is this policy which has been the politically most difficult, has given rise to the greatest problems and the highest drama and is, therefore, the most interesting. The third reason is that I was concerned with the development and management of the conservation policy and it is the policy with which I am most familiar. But, most importantly, the conservation policy is central to the whole of the CFP. Unless what is meant by 'conservation policy' is properly understood and its implications realised, the CFP can never be a success.

After six years of difficult negotiations, the conservation policy was finally agreed on 25 January 1983, a date which is often regarded as that on which the whole CFP was agreed. In fact, the structural and market policies were adopted in 1970 and were being implemented throughout the negotiation of the conservation policy; it is arguable that the increase in the fleets of the Community, subsidized by Community funds, was a major factor in making it so difficult to agree the conservation policy. The start of the policy for international relations cannot be specifically dated because it did not require a 'basic' regulation to be adopted for it to be implemented but it was the extension of fisheries limits to 200 miles at the start of 1977 which gave it its impetus. There was much euphoria on 25 January 1983 because it was expected that the adoption of the conservation policy would result in the efficient management of the fish stocks in Community waters and a prosperous fishing industry. That euphoria has given way to despair and despondency as some of the most important fish stocks have dwindled and the industry has been faced with one major difficulty after another. Ten years after the adoption of the CFP is an appropriate time to evaluate why the expectations of 25 January 1983 remain largely unrealized and what, if anything, went wrong.

One of the major problems in discussing the CFP is a general lack of knowledge about how the Community is structured and operates. Chapter 1 provides this background, without which it is impossible to realize even why there is a CFP at all, let alone the way it is structured and the manner in which it operates. Chapters 2 to 3 outline the gradual evolution of the policy, culminating in the adoption of the conservation policy. These chapters form 'the origin'. Chapters 4 to 9 form the 'evaluation', Chapters 4 and 5 describing how the conservation policy has been operated and Chapter 6 the tortuous decision-making processes of the Community or, as it might be called in respect of the conservation policy, how decision taking is avoided. Chapter 7 evaluates the policy, assessing whether it has been the disastrous failure which so many fishermen believe it to have been, comparing what the Community was trying to negotiate with what it is now thought that it was trying to achieve. Chapter 8 describes basic fisheries science. Although this might be thought inappropriate to a book on the CFP, one of the major reasons for the failure of the conservation policy is the lack of comprehension of the basic principles of fishery science by politicians, fisheries administrators, fishermen and even some scientists, with the consequence that they are unable to communicate effectively; this is made worse by the fact that many who think that they understand do not. This is essential for discussing 'the future'. Chapter 9 examines how the science is applied and describes its limitations. Chapters 10 to 12 concern the future of the CFP. Chapter 10 examines how the decision-making processes might be changed in order that those who know most about the subject, the fishing industry and the fishery scientists, could be involved in a co-operative relationship. The objective should be to have decisions made in a rational manner, after calm deliberation. Chapter 11 examines what the objectives of the CFP should be, concluding that the fisheries of the Community should be managed in order to maximize the economic benefits. How this objective might be achieved is described in Chapter 12.

One of the problems on writing about an on-going policy is that it is continually evolving and, inevitably, what is written today is outdated by decisions taken tomorrow. I completed this book shortly after the 1992 review of the conservation policy, the major development in the history of the CFP since the adoption of the conservation policy of 1983. The book covers this review up to the point of the adoption of Regulation No. 3760/92 establishing a Community system for fisheries and aquaculture which replaced the regulation adopted in 1983. However, I have only been able to speculate on the implications of the provisions of this new regulation for the future development of the CFP. The Commission has also made a proposal for a new regulation concerning control and enforcement. This regulation could have significant implications for the policy but I have been unable to examine these because they will depend upon what provisions the Council eventually adopts. That the policy is on-going also presents the problem of choosing the point in time from which to write; I have chosen as my 'now' February 1993.

This book is written from the point of view of someone who was trained as a fisheries scientist but who had to become part administrator and part lawyer in order to draft legislation based upon the scientific advice. This gave me an insight into the problems which was perhaps unique and which is reflected in this book.

This book will be primarily of interest to those who have to concern themselves with the CFP but it will also be of interest to those who want to understand how the Community operates in developing and operating its policies, a subject of topical interest in the context of the debate on 'Maastricht', 'subsidiarity' and the future of the Community. For this reason the book has been written in a style which, it is hoped, will make it readable to a wide audience, including those outside fisheries. Scientific jargon has been kept to the absolute minimum and I have risked offending my former scientific colleagues by using some terms loosely when I thought that there was more to gain than to lose by doing so. For example, I have referred to 'fishing effort' where, strictly, I should have used 'fishing mortality rate'. For the same reason references have been kept to a minimum and I have deliberately refrained from describing in detail the multitude of fisheries regulations adopted by the Community and the reasons for their adoption, except when this is essential. Citations from regulations have been made only where it was essential to convey precisely what the legislation states.

Much public odium is directed at the 'Eurocrats' living in their ivory tower in Brussels who are accused of trying to run fisheries while knowing nothing about them. For 11 years I was one of those Eurocrats, but this book is not an apologia. It explains what actually happens and how difficult, if not impossible, it is to get the 'right' decisions taken. The reason for which I became involved with the CFP was that throughout the negotiations which had taken place over the accession of the UK to the EEC I had become increasingly aware that responsibility for fisheries would lay more and more with the EEC. At the time I was working at the Fisheries Laboratory, Lowestoft, but was becoming increasingly more interested in trying to apply the results of fisheries research rather than doing it. Therefore, when I was approached to take up a post in the Directorate General for Fisheries (DG XIV), it appealed to me and in 1979 I joined the 'Internal Resources' division of DG XIV. I have, at times, been called 'the architect of the CFP', a phrase which has usually not been intended as a compliment but as the base from which to launch an attack on me for designing such a monstrosity. I make no such claim; when I joined the Commission services much work had already been done on the conservation policy and, in particular, the general principles of access to Community waters and TACs and quotas had been decided. The detailed negotiations had yet to be started and soon after my arrival they commenced in earnest. I rapidly became heavily involved with the development of the conservation policy, in particular with the work on TACs but, as described in the book, the manner in which policy is decided in the Community made me, in theatrical terms, but 'a bit player'. Architects there may have been but the decision-making process meant that the architects' plans were at the mercy of the customers, the Member States, whose modifications negated many plans. In 1986, I was promoted to Head of the 'Conservation' unit where I was responsible for advising the four commissioners whom I served on managing the policy. In that capacity I argued for the policy to be modified so that it might achieve some of the objectives apparently desired by the fishing industry, but without success, for the reasons described. Eventually, after being made Advisor to the Director General in mid-1990, I retired from the Commission at the end of September 1990.

I should like to pay tribute to the many colleagues both within the Commission and in the fisheries departments of the Member States with whom it was my pleasure to work, even if the pleasure was one of arguing opposing points of view. Within DG XIV I owe a debt of gratitude to those who helped me find my feet when I arrived as a politically naive scientist, to those who gave me the opportunity to be so closely associated with the development of the conservation policy and to those who assisted me so ably in the day-to-day struggle of getting regulations drafted against almost impossibly short deadlines. They know who they are. They have my gratitude.

Mike Holden
February 1993

Mike Holden died suddenly in December 1995 when he was about to update the first edition of his book. It is therefore a privilege for me to add the summary of events (Chapter 14) on his behalf, and to comment on future prospects both as a recollection of our long personal association and as Chairman of the Buckland Foundation which appointed Mike as its Professor for 1991.

David Garrod

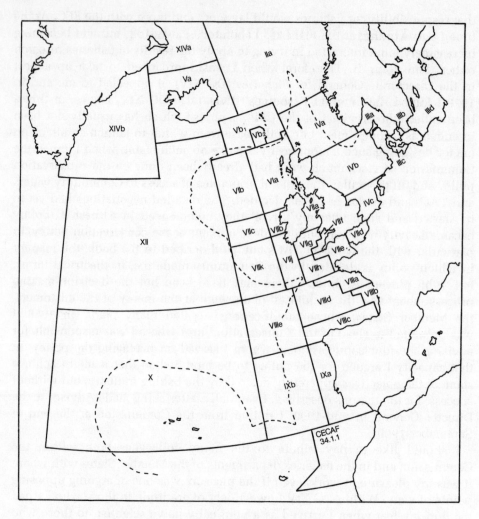

Map of ICES and Fao Eastern Central Atlantic (CECAF) areas showing regions and Community waters; the 200-mile limits are provisional and have no legal status.

Chapter 1

Council, Commission, Court and Parliament

or How The Community Operates

Popular perceptions

It is impossible to understand the operation of the Common Fisheries Policy (CFP) without knowing the legal basis of the Community and the political background to decision-taking in the Community. These are described briefly in this chapter. If the popular press of whatever nationality were to be believed, decisions in the Community are taken by a small group of power-hungry, unelected bureaucrats, the Commission, who live in an ivory tower in Brussels, who consult no-one and are responsible to no-one and that it is only the vigilance of national governments which keeps them in check. Fishermen consider that 'their' industry is run by faceless Brussels bureaucrats who know nothing about fishing yet dictate to them what they can and cannot do (Fig. 1.1).

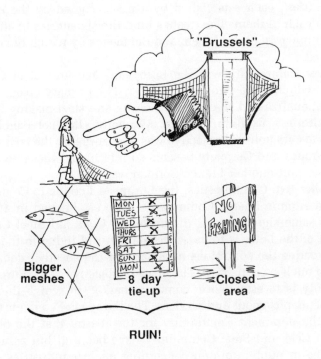

Fig 1.1 The view of the fishing industry of 'Brussels'.

That this belief is widely held is not surprising. Only a minority of people in the Community are involved in its decision-making processes and even fewer attend the meetings of the Council of Ministers where decisions are taken. The discussions on which these decisions are based take place behind closed doors and are not reported in any official document, such as parliamentary records, the minutes of the Council being secret. All the media carries are the reports of the meetings by the ministers concerned, whose accounts are designed to cast themselves in the best light and the Commission in the worst. So to what extent does this popular view of the EEC as 'a law unto itself' represent the truth and to what extent is it a total travesty of the actual situation?

Although the Community is not a perfect system, having been invented by man, this popular image is far from the truth. Like all good systems of government, the treaties establishing the Communities provide a system of checks and balances which are, in many respects, probably more effective than those found in many democracies, where the elected government is responsible for initiating legislation and uses its majority to ensure that the legislation is adopted, even if there is widespread opposition. One has only to think of the adoption in the UK of the Treaty of Maastricht.

The Communities and the treaties

Although it is commonplace to refer to 'the Community', there are three 'Communities', each established by a treaty signed by the countries which agreed to each of them. The treaties bind the signatories to achieve the objectives laid down in each of them and the means by which the objectives are to be realized.

The first Community was the European Coal and Steel Community, the treaty for which came into force in 1951. The primary objective of this treaty was to rationalize and co-ordinate the iron and steel-making industries of the six signatories, Belgium, France, Germany, Italy, Luxembourg and the Netherlands. Its political objective was to ensure that the coal and steel industries of France and Germany became so interlinked that they could never be used to support another Franco-German war.

The other two Communities, the European Economic Community and the European Atomic Energy Community, were established in 1957, again each with the same signatories as the European Coal and Steel Community. The objective of the latter is to 'raise the standard of living in the Member States . . . by creating the conditions necessary for the speedy establishment of and growth of nuclear industries'. The primary objective of the European Economic Community is to establish a 'common market' in which there is free interchange of people, goods and services. The European Economic Community has become the dominant Community for the reason that the objectives of the European Coal and Steel Community have been all but achieved and, with the change in public opinion concerning nuclear industries since 1957, the European Atomic Energy Community has never had a high profile. Thus, although the term 'Community' is a misnomer, it has become so currently accepted that it is used in this book.

The countries which belong to the Communities are called the Member States. The Member States which negotiated the treaties originally had the advantage that they were able to influence their provisions. Countries which join later – accede to the treaties – not only have to accept the terms of the treaties as they stand but also all the legislation based upon them which has been adopted prior to their accession. The whole body of legislation is termed the *acquis communitaire*. A new Member State usually obtains exemptions, known in Community jargon as 'derogations', from those parts of this legislation which it finds it is unable to implement immediately, but derogations are only temporary. Sooner or later they expire. This situation also works in reverse. A new Member State may not benefit from all the advantages of Community policy. For example, both Spain and Portugal had to accept restrictions on their fishing activities on joining the European Economic Community because the existing Member States considered that to allow Spain and Portugal the full provisions of the CFP would have wrecked it.

If the Member States agree the treaties can be modified. Thus the treaty which established the European Economic Community, the Treaty of Rome (named after the city in which it was signed, as is customary), was modified by the Single European Act in 1986. This Act bound the Member States to achieve the Common Market by the end of 1992, progress to that end having been so slow since 1957. The Treaty of Rome will be further modified if and when the Treaty on European Union signed in Maastricht in February 1992 is ratified by all the Member States. This treaty sets out the conditions and timetable for implementing one of the objectives of the Treaty of Rome – 'approximating the economic policies of Member States'.

As the CFP was established under the provisions of the Treaty of Rome, in this book the term 'Community' always refers to the EEC and 'the Treaty' to the Treaty of Rome. Upon ratification of the Treaty of Maastricht, the European Economic Community will legally become the European Community, a term which is already in general use.

The institutions of the Community

The Community consists of four institutions, the Council, the Commission, the European Parliament and the Court of Justice. These institutions have been common to the three Communities since 1965. As with all the fundamental rules of the EEC, it is the Treaty which provides for these institutions and specifies their powers.

The Council of the European Communities

The Council is considered the primary institution because only it has the power to adopt legislation, although it can delegate this power to the

Commission. The Council consists of representatives of the Member States. It is served by its own secretariat of 2215 civil servants (in 1992) who are employed directly by the Council, not by the Member States. All meetings are presided over by the presidency which rotates between the Member States on a six monthly basis in alphabetical order determined by the spelling of the country in its own language, as given in the list showing the number of votes held by each Member State.

The system of voting in the Council is that of qualified majority voting except where the Treaty specifies unanimity, which is nowhere the case in fisheries. The number of votes held by each Member State is as follows:

Belgium	5
Denmark	3
Germany	10
Greece	5
Spain	8
France	10
Ireland	3
Italy	10
Luxembourg	2
Netherlands	5
Portugal	5
UK	10

making a total of 76 votes. A qualified majority is 54 votes when voting on a proposal by the Commission and 54 votes cast by at least eight Member States in certain specific cases specified by the Treaty, which are not applicable to fisheries. The system of qualified majority voting means that there must be a minimum of seven Member States in favour of a proposal. Twenty-three votes are sufficient to prevent adoption of a proposal, for which reason it is termed 'a blocking minority'. This is one of the most important of the 'checks and balances' mentioned earlier. It means that there must always be widespread agreement amongst the Member States before a proposal for legislation can be adopted while avoiding the need for unanimity, which would give each Member State an effective veto. The number of votes held is biassed in favour of the smaller Member States, which further prevents their interests being overridden by those of the larger Member States.

The powers of the Council are:

- to take decisions;
- to co-ordinate the economic policies of the Member States; and
- delegate responsibilities to the Commission.

The Council cannot initiate proposals for legislation, only examine and vote upon proposals made by the Commission. It also cannot amend a proposal of the Commission except by unanimity, which has occurred only once in fisheries. The Commission can withdraw a proposal at any stage before it has been voted upon by the Council. This is also an important part of the system of 'checks and balances'.

The Commission of the European Communities

The term 'Commission' is that most widely misused in Community language because it is sometimes used to mean the commissioners, sometimes the institution and sometimes the civil servants who work for the institution. Formally, as defined by the Treaty, the Commission consists of the commissioners. Commissioners are appointed by the Member States but on appointment take an oath pledging themselves to work only in the interests of the Communities and act independently of all national interest, which unfortunately is not always the case. There are 17 commissioners with a minimum of one but no more than two from each Member State; in practice this means that the largest Member States, France, Germany, Italy, Spain and the UK each have two and the smaller Member States one. From the commissioners are appointed, again by the Member States, a president and six vice-presidents but they do not have any more powers than their fellow commissioners and for this reason the commissioners acting together are often referred to as 'the college' which is a useful term to make it explicit when the term 'Commission' is being used in its strictly formal sense to refer to the commissioners acting together. The term of office of the commissioners is four years but becomes five years upon ratification of the Treaty of Maastricht. The president and vice-presidents are appointed for a term of two years; if not reappointed, they revert to being a commissioner, unless the whole Commission is due for renewal. This provision is also modified by the Treaty of Maastricht whose Article 11 provides that 'The Commission may appoint a Vice-President or two Vice-Presidents from among its members'. The objective of changing the term of office of the Commission is to make both the length of the term and its period coincide with that of the European Parliament. To that end the term of office of the commissioners appointed in 1993 was only two years.

Commissioners cannot be dismissed except for serious misconduct, although the whole Commission can be dismissed by the European Parliament, neither event ever having happened. Decisions are made by the college by simple majority voting, each commissioner having one vote. Commissioners are equivalent to ministers in national governments, each being responsible for specific policies or areas of Commission administration.

In brief, the powers of the Commission are:

- to initiate proposals for legislation;
- to ensure that the provisions of the Treaty are carried out;
- to exercise powers delegated to it by the Council; and
- to formulate recommendations and deliver opinions.

The Commission employs its own civil servants who are often termed 'the services', which is useful in making the distinction between two of the senses in which the term 'the Commission' is used. Again, it is a common misconception that 'the services' are national civil servants whose task is to represent the interests of their Member State. Although many, but by no means all, are recruited from national civil services, the civil servants working in the Commission owe their allegiance to that institution. Admittedly it is impossible to leave behind all national attitudes and prejudices on becoming an employee of the Commission but the fact that the administrative units consist of a mixture of nationalities militates against national attitudes being allowed to influence policies.

The European Parliament

The European Parliament consists of 567 members elected by universal suffrage. The number of members was increased at the European Summit held in Edinburgh in December 1992 in order to take account of the reunification of Germany, the only institution whose numbers were changed. Their allocation by Member State is as follows:

Belgium	25	Ireland	15
Denmark	16	Italy	87
Germany	99	Luxembourg	6
Greece	25	Netherlands	31
Spain	64	Portugal	25
France	87	UK	87

The numbers approximately correspond to population size but those Member States having the smallest number of inhabitants have proportionately more than those with the largest. As in the UK, all constituencies do not have the same numbers of voters. MEPs co-operate by political parties, not by nationality.

The European Parliament has very limited powers, the most important being that it is co-responsible with the Council for the Community's budget and may dismiss the Commission. Where specified in the Treaty its opinion has to be obtained on proposals for legislation, in which case there exists a procedure for co-operation between the Council and the European Parliament laid down in the Single European Act. The European Parliament has a sub-committee on fisheries, part of its Agricultural Committee, before which the commissioner and representatives of 'the services' appear to explain the policy and proposals of the Commission. But only if a proposal for legislation is based upon Article 43 of the Treaty, which stipulates for fisheries that the European Parliament must be consulted, can it give an opinion and the co-operation procedure between the Parliament and Council have to be followed before a proposal can be adopted by the Council.

This lack of power, which is often referred to as 'the democratic deficit' of the Community, is bitterly resented by the European Parliament which seeks to obtain more, a subject presently very much under debate within the Community but outside the remit of this book. It is certainly arguable that the European Parliament, which is the only institution whose members are democratically elected, should have more powers. Be that as it may, in the present situation, the European Parliament does not have a commitment to getting the legislation proposed by the Commission adopted as if it would have if it formed a government. This has two consequences. First, each member of the European Parliament (MEP) argues for the interests of their constituents. As one of the major problems of managing fisheries is trying to find solutions which reconcile often irreconcilable positions, having MEPs – each of whom is arguing in favour of one particular position – is not helpful in arriving at solutions. Second, the European Parliament shows no urgency in reaching decisions. As much fisheries legislation is concerned with day-to-day management, consultation would cause serious delays; the provisions on cooperation between the Council and the European Parliament contained in the Single European Act allow for a period of up to five months in which to carry out the cooperation procedure, which does not include obtaining the Opinion of the European Parliament, for which no time limit is given in the Act. It is already a common complaint of the fishing industry that planning is made difficult or impossible because decisions are often taken only shortly before they come into effect; this is particularly the case with adoption of TACs. Even if the European Parliament took decisions rapidly, having to consult it on all matters would impose additional delays.

When the MEPs do make common cause, the European Parliament can be very effective. For example, it was a result of its pressure that budgetary provisions were made for the Community's fisheries inspectors. It has also played an important role in such issues as hunting seals and the use of large gillnets on the high seas.

The European Court of Justice

The Court consists of 13 judges assisted by six advocates-general appointed by the Member States. Their term of office is six years. The duty of the Court of Justice is to ensure that the interpretation and application of the treaties are observed. Cases may be brought by one institution against another, by institutions against Member States and *vice versa* or by individuals against an institution or Member State if any of them considers that Community law has been infringed. The Court also gives decisions concerning interpretation of Community law in cases referred to it by national courts.

Decisions of the Court often have a major impact on the development of policies, which has been the case in fisheries. Six judgements of the Court made in the period prior to the settlement of the CFP confirmed that the Community has total responsibility – 'competence', to use euro-jargon – for fisheries. These decisions were as follows. In a judgement made by the Court of Justice in 1976 in a case referred to it by a Dutch local court concerning overfishing of quotas, the Court confirmed the power of the Community to take conservation measures and to enter into international agreements in fisheries. In five subsequent cases brought by the Commission, one concerning a measure adopted by Ireland and four concerning legislation adopted by the UK, the Court of Justice ruled against the two Member States. In the case against Ireland the Court ruled that its measure was discriminatory and in the cases concerning the UK that it had not notified or consulted the Commission before the UK had adopted its measures.

Adopting Community legislation

Community legislation has three legally binding forms:

(1) regulations which are directly applicable and binding in all Member States;

(2) directives, which are also binding in all Member States but leave the detailed legislation to be enacted by each Member State; and

(3) decisions, which are binding upon those to whom they are addressed.

The Council and Commission can also make recommendations and give opinions but neither of these is legally binding. Most legislation concerning fisheries is in the form of regulations.

This section describes how legislation is adopted by the Council. The procedure by which the Commission adopts legislation is described later.

Adopting Council legislation

The role of the Commission
Only the Commission may propose legislation but it cannot make proposals on any subject it chooses; the basis for it must exist in the treaties. But the treaties provide only the broad outlines of each policy; for example, the whole of the CFP, as well as the Common Agricultural Policy, stems from one indent in Article 3 of the Treaty containing 11 words, elaborated in a subsequent article, Article 38, containing 35 words. The detailed implementation of the treaties therefore lies very much within the determination of the Commission to implement a policy and, in particular, the commissioner responsible for each field of policy.

The Commission has a legal responsibility to implement the treaties and advance Community policies and it must and does take the initiative to do this. This is the reason for which the Commission is apparently always trying to extend and enlarge its powers, giving rise to the misunderstanding that the Commission is power-hungry. All that the Commission is doing is exercising its powers under the treaties, which have been agreed by the Member States. Nothing can finally become Community law until adopted by the Council, the Member States acting together, except in specific areas in which the Council has delegated its power to the Commission and these areas concern only matters of little or no political importance.

Making the proposal
Although it is the Commission which must formally make the proposal, the initiative for legislation may come from many sources. That which led to the establishment of the CFP came from France and Italy asking the Commission to make proposals concerning structures and markets. A more recent example is the modification of the structures regulation to permit Community funding of vessels smaller than nine metres, proposed at the instigation of Italy and Greece which considered that the previous minimum limit of nine metres discriminated against their fleets. The policy for international fisheries relations

is also Member State-driven as far as individual agreements are concerned but it is the Commission which has developed the ideas on how this policy should evolve in the face of third countries wanting to exploit their own resources. In the case of the conservation policy it is the scientific advice which almost entirely determines what the Commission proposes. The Commission is also very open to representations from the fishing industry, the reason being that the services do not have the necessary expertise amongst its limited staff, nor the relevant data. Lobbying can be very effective but the extent to which it is carried out varies from Member State to Member State. The UK industry has used this means very little to date, preferring to work through government officials. The emphasis given to the 'Shetland box' in the Commission's Report 1991 on the Common Fisheries Policy shows how effective lobbying can be.

Depending upon the nature of the legislation, there may be extensive consultation with national governments and fishing industries. The Commission frequently convenes working groups of experts to advise it before the initial proposal is drafted. This is much more common in the field of structures than for other policies because structural measures are rarely urgent and usually lay down legislation which covers a long period. In contrast, proposals for regulations in the field of conservation policy are usually concerned with immediate management, allowing little or no time for prior consultation. The reason for which there is so much consultation, if possible, is that the legislation has to be adopted by the Council and there is little point in the Commission making a proposal which, after debate and modification, will not eventually command a qualified majority. It is important for the Commission to get its proposals adopted, even in a modified form because it reflects badly on the political judgment of the Commission if they are not adopted. This does not prevent the Commission from 'testing the waters' by making proposals which it knows will not be accepted immediately or only after several attempts. Also, what may be regarded as unacceptable proposals may be included as a negotiating ploy, the Commission deliberately including them knowing that they will have to be dropped or amended eventually but as part of a compromise permitting the adoption of other measures which the Member States would have otherwise not adopted.

Proposals for legislation are drafted by the services. Once the text has been drafted it has to be translated into all nine official Community languages and these texts examined to ensure that they mean the same thing in all nine languages; despite this, they sometimes do not. The draft proposal is then submitted to the 'college' which examines and votes upon it, but often not until after much heated discussion and debate amongst the commissioners over its contents, which may be considerably revised before it is adopted. Following its adoption, the proposal is forwarded to the Council (Fig. 1.2).

Adopting legislation
It is rarely that a proposal commands a qualified majority without debate. Perhaps the only occasion on which this occurs is a proposal to increase a TAC and only then if the Council considers that there is no possibility of getting the Commission to propose more. Usually a proposal contains much that one

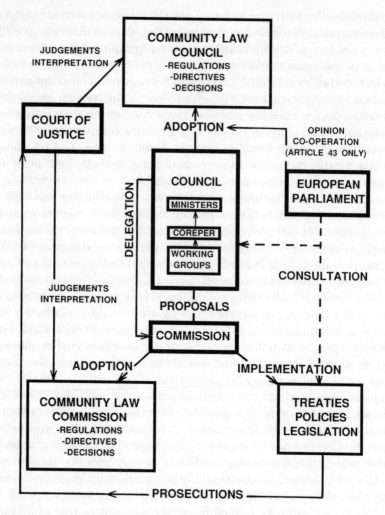

Fig. 1.2 The institutions of the Community, the relationship between them and their main powers.

or more Member States wants modified or omitted. A proposal of the Commission is, therefore, not a 'take it or leave it' affair. It is expected that much discussion will take place, particularly immediately prior to a meeting of ministers, in order to find the basis for a compromise solution which can be adopted. A proposal which has reached the stage of being discussed by the ministers will also contain many elements which the Member States will want adopted, but there will also be others, different for each Member State, which they will not want adopted or adopted in an amended form, if they can possibly negotiate it. The position of each Member State will also be a compromise reflecting the fact that the fishing industry of each is very fragmented with each sector having often widely different and conflicting objectives. A proposal is, therefore, subject to much examination and invariably the original proposal is much amended before being finally adopted.

The Council operates at three levels, of which the highest is that of ministers. Member States are represented at each Council meeting by the national ministers responsible for the policy to be discussed. Only ministers can take decisions which are legally binding. The next highest level is that of ambassadors of the Member States to the Community, the Committee of Permanent Representatives (COREPER). Lastly, there is the level of working groups which consist mainly of national civil servants working in the area of policy in their home departments who travel to Brussels for relevant meetings. The Commission is represented at meetings at all three levels, by the appropriate commissioner at meetings of ministers, usually by the Director General or a Director at COREPER and by the representatives from the unit responsible for drafting the proposal at the working groups.

The examination of a proposal of the Commission starts in the relevant working group, of which there are two in fisheries – the External Working Group, which discusses all matters concerning relations with third countries, including examination of the relevant proposals for legislation, and the Internal Working Group which deals with policies for conservation, markets and structures. Different representatives, having the relevant expertise, attend the meetings of the Internal Working Group. In examining a proposal for legislation the objectives of a working group are (a) to provide the opportunity for the representative of the Commission to explain the reasons for which the Commission is making its proposal, (b) to permit the representatives of the Member States to state their position regarding the proposal and to make comments and suggestions, (c) to rectify problems of drafting – not only may the proposal be badly drafted but working in nine languages gives plenty of opportunities for mistranslation – and, most importantly, (d) to identify the main political issues which need to be examined at the next level, COREPER. Depending upon the complexity and urgency of the proposal, there may be several meetings of a working group spread over many months. In the intervals between meetings both the Commission and the Member States consider their positions and, if thought necessary, the Commission may amend its proposal.

Once the working group has made as much progress as possible, the proposal is examined by COREPER which has a certain freedom of political initiative not possessed by the working groups. If possible, COREPER tries to achieve agreement in order to avoid ministers having to discuss the proposal, the objective being to leave for ministerial decision only those matters of major political importance. However, even if COREPER reaches agreement, the proposal still has to be formally approved at a meeting of ministers but this can be done at any Council meeting of ministers, where it is adopted without further discussion as what is known as an 'A' point. Since qualified majority voting was re-established, it is becoming more frequent for COREPER to be able to reach agreement. If a Member State realises that it is going to be in a minority, it will often agree to a proposal being adopted, subject to its position being recorded in the minutes of the Council.

If agreement cannot be reached in COREPER, the proposal is examined by ministers. The Presidency and the Commission, usually working together, will try to find a way of reconciling positions in order that a qualified majority can be obtained. It is important for the Presidency to achieve agreement for two reasons. First, its political prestige is at stake. Second, it determines the agenda for the six months for which it holds the presidency, during which it is in a very strong position to ensure that proposals in which it is interested are adopted; conversely, it can ensure that those in which it is not interested or to which it is opposed are shelved. The magic number, of course, is 54 votes, to obtain which concessions are often made to Member States to ensure they vote for the proposal. Eventually, when the Presidency and the Commission consider that they have agreement, the Commission drafts its final compromise proposal, on which the Council votes. If a proposal receives the necessary qualified majority, it becomes law and is published in the Official Journal of the European Communities. If it does not, it remains on the table, unless withdrawn by the Commission, and the whole process is restarted, sometimes based on an unchanged proposal but more often based on a modified proposal.

Although there is a formal procedure of working groups, COREPER and the Council, the final decisions taken depend upon an informal process of lobbying and negotiation – 'horse trading', to put it bluntly. As already stated, the contents of the initial proposal depend upon the pressure placed upon the Commission, the information available to it and the directions given to the services by the commissioner who may, or may not have, very strong views about the policy he or she wishes to be pursued. The twelfth modification to the technical conservation measures adopted in January 1992 stemmed very much from the strong initiatives taken by the commissioner then responsible for fisheries, Vice-President Marin.

The first round of horse trading usually occurs even before the proposal drafted by the services is adopted by the Commission. Despite the fact that commissioners are duty bound to 'neither seek nor take instructions from any Government or from any other body' draft proposals are invariably leaked to the Member States which then seek to have the draft modified before it is adopted by the Commission. It is easier to get modifications done in a series of 'deals' at this level than it may be later in the open debate in the Council.

Little or no horse trading takes place in the working groups because the representatives from the Commission and the Member States are given little or no powers by their authorities to alter the draft; the representatives of the Member State ask for clarifications and state their positions, which are noted by that of the Commission who also responds to points of clarification. The overriding reason for this is that the fishing industry regards it as a sacrifice of the national position if any concessions are made at this level and no minister can accept being accused of selling out his industry; middle-rank civil servants are considered to be mere technicians; decisions must be taken by politicians.

The members of COREPER are delegated to take decisions within the remit of their negotiating briefs and this is probably the most effective level of negotiation, not only for this reason but also because most members of COREPER have spent a long time in their posts, are familiar with the subject and are well known to each other. Above all, they are all multi-lingual and the need for interpretation is minimal. Working in the Community is a constant reminder of the Tower of Babel.

It is immediately prior to and during a meeting of the Council of Ministers that the real horse trading starts. Prior to the meeting there will be many contacts. These may take place at many levels depending upon who it is thought can wield influence. Thus, the civil servant from the national permanent representation, which each Member State has in Brussels, may contact the head of the unit responsible for preparing the legislation or his director to try to determine what the position of the Commission might be during the Council meeting and influence the shape of any compromise. More importantly, the Director General will either meet with his opposite numbers in Brussels or in their capitals in order to try to determine which issues are important to them, which they are determined to get adopted and where the possibilities of a compromise exist. In the light of this exchange of views it will be decided which demands of which Member States are not subject to compromise, whose positions will be favoured in order to gain a qualified majority and whose will be sacrificed. If no obvious qualified majority is evident, the Commission will decide what concessions will have to be made to which Member States in order to obtain one. On the basis of this evaluation, the Commission will prepare in advance the outlines of what it expects to be the final compromise.

A typical meeting of the Council starts with a *tour de table* in which each minister sets out his/her position, often usually reiterating little or no more than has been said by his/her civil servants at meetings of the working group. The meeting will then adjourn to allow for meetings between the President, the commissioner and each minister, each accompanied by a few senior officials. These meetings are known as 'confessionals' from their resemblance to the Roman Catholic practice of confession. This practice partly explains why Council meetings take so long; a meeting of 15–20 minutes with each Member State means a total of something like four hours or more, given that there are 11 interested Member States; at least in fisheries Luxembourg has no interests. Following this stage difficult negotiations on final compromises or the wording of the final drafting of articles may continue for several hours. At the end of this period what it is hoped a final text will be prepared. Even though much of this might have been drafted beforehand, the final document has to be typed, translated into all nine languages and photocopied before it can be distributed to ministers, giving rise to further delays. Usually, this text will be presented as that of the Presidency and the Commission, in which case it represents a formal proposal of the Commission. If the Presidency and Commission have done their work well the compromise may be adopted; this is never a formality. Each minister feels honour-bound to state that he does not agree with it but will accept it only at great sacrifice and only if all other ministers accept it as it stands. This means that only one minister has to state that he cannot accept it for the whole debate to reopen. Often at this stage the debate may be about what are, on the face of it, relatively trivial issues, such as a few tonnes of fish to be added to a TAC. But if this few tonnes were conceded every minister would demand his 'pound of flesh'. Although it is only conjecture, the limit of 135 days at sea in 1992 for vessels fishing in the North Sea and west of Scotland for cod and haddock probably gave rise to such a discussion, with various figures being vigorously debated before agreement was reached.

Different techniques are used to obtain a qualified majority. A much used technique is for the Presidency to present the text as one which has been carefully balanced to take account of the major interests of all Member States, although admittedly containing advantages and disadvantages for each. In order to increase the chances of getting a qualified majority, the compromise text may often include legislation concerning all four policies, thus providing a mixture of potential gains for all Member States, which has to be accepted or rejected in its entirety. In this situation voting will always start with the Member State which holds the Presidency, whose vote will automatically be cast in favour. Invariably though the minister hedges his vote with the provisos described above, thus providing an escape route if some Member States do not vote in favour but also politely blackmailing the other Member States to do so. Voting then continues with all holding their breath while counting the votes. If a qualified majority has been obtained, there is a deep sigh of relief. If the vote fails the meeting may end or it may be back to the drawing board for another round of negotiation. Often in this situation the Council will enter into restricted session, with only ministers and a few senior officials present, or even a 'super-restricted' session in which only ministers and their most senior official from each delegation are present, together with the secretariat of the Council.

Exceptionally, the Presidency presents a compromise text with which the Commission does not agree, in which case it does not constitute a formal proposal. The situation then depends upon subsequent events. If all the Member States were to agree the text, the Council would in effect modify the original proposal of the Commission by unanimity and it would be adopted. To my knowledge this has never happened in fisheries for the reason that, if the Commission is not prepared to be co-responsible for a text, it is certain that the text will not be accepted by unanimity or even command a qualified majority. Continuing this scenario, the subsequent debate may result in a modified text which commands a qualified majority and which the Commission is then prepared to make a formal proposal. The more usual outcome in this situation is that the compromise text is not adopted and that a new one is presented by the Commission based upon the debate on the text produced by the Presidency.

However agreement is reached, invariably the final legislation is accompanied by declarations of the Presidency, the Commission and usually some Member States. These declarations have no legal status. They simply clarify the position of their originators, whether it is to work to agreement on some specific issue within a stated deadline or to state how a particular Member State interprets the legislation.

Adopting Commission legislation; the Management Committee procedure

The Commission may adopt legislation only for areas of policy for which the Council has delegated responsibility. These never include politically important matters. Usually they concern day-to-day management of a policy, often of a technically detailed nature, where rapid decision taking is required. In fisheries, many aspects of the implementation of structural and market policies are implemented by Commission regulations. In the conservation policy, examples of Commission regulations are those concerning measuring of mesh sizes, attachments to nets and sampling of by-catches.

The process of drafting such legislation is similar to that described for proposed Council legislation but once it has been drafted in most cases, and always in fisheries, it is submitted to a 'Management Committee' established for each area of policy. The composition of management committees is identical to that of working groups with representatives from all Member States but, unlike the situation described for the Council, the Commission presides and can call a vote at any time. The voting system is identical except that the committee is called upon to give 'an opinion' by which the Commission is not bound. Even if the proposed regulation receives a negative opinion, that is, there is a qualified majority of 54 votes against it, the Commission can still adopt it but in this case the Council has the right to discuss and vote upon it. In the meantime the legislation takes effect. In the cases that there is a qualified majority in favour of the proposal or no majority at all the Commission may proceed with the adoption of the proposal.

The hierarchy of legislation

The system of Community legislation forms a hierarchy. This can be observed in any Community regulation which, in its preamble, states the article or articles in the preceding legislation on which the regulation is based. The primary legislation is the Treaty, which is usually very concise, as already noted.

Secondary legislation is usually in the form of regulations which expand upon the principles in the treaties to lay down the basic framework of a particular policy; for this reason, they are termed 'basic regulations'. For example, Council Regulation (EEC) No. 170/83 of 25 January 1983 establishing a Community system for the conservation of fishery resources was the basic regulation establishing the conservation policy and was based upon Article 43 of the Treaty. (It has now been replaced by Council Regulation No. 3760/92 of 20 December 1992.) Tertiary legislation lays down the details of how the policy is to be implemented; for example, regulations concerning TACs and technical conservation measures were based upon Regulation No. 170/83 (and will in future be based on Regulation No. 3760/92). Commission regulations, adopted under powers delegated by the Council forms a fourth level and is based upon the relevant article in the tertiary regulations.

The legal basis for a regulation has one practical consequence of great significance. As already noted, regulations based upon Article 43 of the Treaty must be submitted to the European Parliament whereas other proposals do not have to be submitted unless the basic regulation makes specific provision for this.

Chapter 2

The Policies for Structures, Markets and External Fisheries

or How the Policy Started

The background

Why a Common Fisheries Policy?

It is frequently queried why there should be a Common Fisheries Policy. As already explained in Chapter 1, the areas in which the Member States have agreed to devolve power to the Communities are laid down in the treaties and the basis for the CFP is to be found in Article 3 of the Treaty which states:

'For the purposes set out in Article 2, the activities of the Community shall include . . . (d) the adoption of a common policy in the sphere of agriculture . . .'

and in Article 38 of the Treaty whose first paragraph states:

'The common market shall extend to agriculture and trade in agricultural products. 'Agricultural products' means the products of the soil, of stock-farming and of fisheries and products of first-stage processing directly related to these products.'

In signing the Treaty, the original six Member States agreed that there would be a common policy for agriculture, which was defined as including fisheries. This provides the legal basis for the policy. All the countries which acceded to the EEC after 1957 have had to accept this provision because of the application of the rule of accepting the *acquis communitaire*.

It is possible to speculate whether it was a sheer accident that fisheries was included in the article. The six original Member States of the EEC, Belgium, Denmark, France, Germany, Italy and Luxembourg were primarily concerned with agriculture, not fisheries. The drafters of the Treaty may have included fisheries only because agriculture and fisheries are often associated, as in the UK, where they are the responsibility of the same ministry. If they had not done so there would not be a CFP today. Be that as it may, once the Treaty as drafted had been ratified by all the Member States, the Commission was under a legal obligation to propose the CFP.

The wherefores of a Common Fisheries Policy

There is a world of difference between including a policy in the Treaty and actually adopting and implementing it; for example, the Treaty provides for a

16

common transport policy but one has yet to be agreed, 35 years later. More generally, the overall objective of the Treaty is to establish a common market yet the 'Single Market' came into existence only at the end of 1992. Even now, it is far from being the Single Market which the founders of the Community envisaged. So why did a CFP get adopted relatively so rapidly, even if the number of years which it eventually took to complete it might be considered a long time, and why are its components those which exist rather than something different?

The link between agriculture and fisheries as well as the overriding interest of the original six Member States in agriculture has already been mentioned. Furthermore, in the early days of the Community, fisheries was the responsibility of a division, one of the Commission's smallest administrative units, in the Directorate General for Agriculture. Fisheries did not become a Directorate General in its own right until 1976, on what many might think was the ironic date of 1 April, known appropriately in French as *poisson d'avril*.

In 1957 the Directorate General for Agriculture was totally involved with developing the Common Agricultural Policy and had no resources to spare for developing a Common Fisheries Policy. It is so commonplace for the media to refer to the 'vast bureaucracy in Brussels' that it is rarely realised that the total permanent staff of the Commission even today numbers under 13 000, of which approximately 40% are involved with translation. This compares with over half a million civil servants in the UK alone and is less than the total number of staff employed by a single ministry or department or a large local authority in the UK. To put it further into perspective, the 'Conservation' unit for which I was responsible, consisted in total of a maximum of ten people, of which five, including myself, were scientifically qualified. Not only does the unit have to keep abreast of all the scientific work published on fisheries in all Community waters in order to advise the commissioner responsible for fisheries and draft the appropriate legislation but it has to represent the Commission at the many Council meetings described in Chapter 1 and assist in negotiations with third countries. At meetings of the Council's Internal Working Group the UK delegation often consisted of more people than scientifically qualified staff in the Conservation Unit. Additionally, the UK delegation consisted of the tip of the iceberg, there being a whole section of the ministry plus two research laboratories to provide back-up; the Conservation Unit, on the other hand, had none. One of the reasons for which there are not more staff is that the Commission does not decide how many people it may employ. The Council and the European Parliament are the budgetary authorities which decide staffing levels and do not allow the creation of additional posts until the level of the workload demands it. As there was no interest in a fisheries policy in the late 1950s and early 1960s, no posts were made available. (This budgetary responsibility for posts can be used by the Member States to prevent the Commission rapidly developing policies in which they are not interested or to which they may even be opposed.)

A second factor making for little interest in developing a CFP was that, in the 1950s, almost 90% of the the catches of Belgium, France, Germany, Italy and the Netherlands were taken outside their national fishery limits, which extended in general only to three miles; Luxembourg, the sixth Member State, is landlocked and had no interest at all in a fisheries policy.

However, this was to change as the implementation of the common market provisions began to have their effects on fisheries, although indirectly, through the Common Customs Tariff. This was more liberal than that which had existed hitherto in France and Italy and its operation began to have adverse effects upon their fisheries interests. They therefore requested the Commission to make proposals for regulations by which assistance could be provided to help them modernize their fleets and infrastructure in order that they could compete. This is a prime example of the initiative for legislation coming from the Member States.

The Commission seized this opportunity to produce in 1967 a long paper, entitled 'Basic principles for a common fisheries policy'. The paper was concerned mainly with structural matters, markets, external trade and social questions. Conservation was not even considered to be a major issue; only one page out of twenty was devoted to the subject and this in the section on structural policy under the heading 'Production'. It is not surprising that the paper addressed itself primarily to these subjects. France and Italy had specifically requested structural and market provisions, similar to those in the field of agriculture with which the Commission was very familiar. At this stage, with fisheries being handled by a division within the Directorate General for Agriculture, there was no expertise in fisheries. That the paper also addressed social questions probably reflected the strong French interest in such matters. However, these three topics and especially structures and markets were routine and the paper broke no new ground in this area, proposing to make provisions under the European Agricultural Guidance and Guarantee Fund (FEOGA, the acronym being derived from the French) for funds to be made available for restructuring the fishing industry ('Guidance') and supporting the markets ('Guarantee'). In contrast, the ideas on 'Production' broke new ground, setting a course which was to be followed until the conclusion of the CFP on 25 January 1983, which is often regarded as the date on which the CFP came into being. In fact, this was the date on which the CFP was completed with the adoption of the regulations establishing the conservation policy. By then the structural and markets policies had already been in operation for 13 years. The evolution of the CFP was a very long process which took, in total, about 15 years, with the conservation policy itself taking about seven years' hard and difficult negotiation. It is impossible to summarize all the discussions which took place in hundreds of meetings in one chapter of one book and no attempt is made to do so. These are described elsewhere in other books specifically devoted to this subject, cited in the references.

The beginnings of the Common Fisheries Policy

Despite the interest of France and Italy in having structural and market policies for fisheries, the Council made little progress in acting upon the paper of the Commission and its proposals based upon it, the reason being that the Member States were divided upon the issue. Germany in particular, being the major contributor to the Community budget, saw no reason why it should effectively provide funds to support its competitors. Germany was supported by Belgium and the Netherlands, all three having efficient fishing industries which did not need financial support.

What broke this impasse was the application by Denmark, Norway, Ireland and the UK to join the Community. All four applicants had major fisheries interests and much of the fishing by three of the original six Member States took place in their waters. If the 'six' did not agree a policy before the applicants joined, the new Member States would be able to influence its development, in particular the conditions of fishing and access to national waters. With a policy in place prior to their accession, the new Member States would have to accept the *acquis communitaire*. The 'six' just managed to agree, almost literally at the eleventh hour, adopting two regulations, Council Regulation (EEC) No. 2142/70 on the common organization of the market in fishery products and Council Regulation (EEC) No. 2141/70 laying down a common structural policy for the fishing industry. Much to the fury of the acceding Member States, although these regulations were adopted by the Council just hours before the negotiations with them started, they did not come into force until later and therefore did not count as *acquis communitaire* in their eyes. However, the original 'six' claimed that they did, a position which they were able to maintain successfully. The most contentious part of these regulations were those in paragraph 1 of Article 2 of Regulation No. 2141/70, since replaced by Regulation No. 101/76, which states:

'1. Rules applied by each Member State in respect of fishing in the maritime waters coming under its sovereignty or within its jurisdiction shall not lead to differences in treatment of other Member States.

Member States shall ensure in particular equal conditions of access to and use of the fishing grounds situated in the waters referred to in the preceding subparagraph for all fishing vessels flying the flag of a Member State and registered in Community territory.'

These provisions meant that no Member State could exclude the vessels of any other Member State from fishing in its waters, a provision which was summarized by the single word 'access' or, more popularly as 'fishing up to the beaches'. Access was to become a major issue in the development of the conservation policy, as described in Chapter 3, made even more difficult to resolve by the bitterness caused by the manner in which these regulations were adopted.

The structural policy

The objectives

The objectives of this policy as described in Regulation No. 101/76 were:

'. . . to promote harmonious and balanced development of this industry within the general economy and to encourage rational use of the biological resources of the sea and of inland waters.'

These objectives have changed little since, except that with each succeeding structural regulation the objectives have been spelt out in more detail. That the original regulation mentions 'inland waters' is interesting; it is usually forgotten that the CFP includes fisheries in inland waters although the Community has never tried to exert its competence there for obvious reasons.

The objectives were described in a little more detail in Regulation No. 2909/83 on a common measure for restructuring, modernizing and developing the fishing industry and for developing aquaculture but it was not until the adoption of Council Regulation (EEC) No. 4028/86 of 18 December 1986 on Community measures to improve and adapt structures in the fisheries and aquaculture sector that the objectives were fully amplified, as follows:

'. . . to facilitate change in the fisheries sector within the guidelines of the common fisheries policy . . . grant Community financial aid for:

(a) the restructuring, renewal and modernization of the fishing fleet;
(b) the development of aquaculture and the establishment of protected marine areas with a view to improved management of inshore fishing grounds;
(c) the reorientation of fishing activities by means of exploratory fishing voyages, redeployment operations, joint ventures and joint enterprises;
(d) the adjustment of fishing capacity by the temporary or permanent withdrawal of vessels from fishing activities;
(e) the provision of facilities at ports for improving the conditions in which products are obtained and landed;
(f) the search for new outlets for surplus for products derived from surplus or underfished species, and for aquaculture products which, by virtue of their rapid growth, pose problems of disposal on the Community market.'

Although France and Italy had tried in 1960 to get social measures included as part of the structural policy, they were not included in Regulation No. 2141/70, probably as part of the compromise which permitted Regulations No. 2141/70 and No. 2142/70 to be adopted. Social measures have never formed part of the structural policy, although with the emphasis of the present policy on contracting the industry there is once more pressure for them to be included, although without success to date.

Throughout 1992 the commissioner responsible for fisheries stated his intention to get a specific fisheries objective added to the European Structural Fund. This fund provides support to regions which are lagging behind economically, those affected by serious industrial decline, and underdeveloped rural areas. Fisheries already benefit to some extent from this fund under its Objectives III and IV, which provide for resolving the problems of long-term unemployment and creating employment for young people. Under what would have become Objective VI, funds would have become available specifically for alleviating the social impact of reducing fishing activities which will result from implementing the multi-annual guidance programmes. Objective VI was also intended to provide funds for improving control and monitoring, giving incentives to use environmentally friendly fishing gear and assisting fishermen's cooperatives. However, the Commission made no proposal and the idea was eventually killed by events at the European Summit held in Edinburgh in December 1992. One of the main bones of contention at this meeting was the question of the extent to which the Community budget should be increased under what was known as the 'Delors II package'. The Summit eventually agreed an increase less than that proposed, thus providing no margin for enlarging the scope of the fund.

The implementation of the policy

The Community has a deficit in supplies of most species of fish, so the obvious primary objective of the policy in 1970 was to eliminate it. The apparently self-evident way in which to achieve this objective was to encourage the building of more vessels to catch more fish. This objective and the means by which to attain it are still held valid by many, particularly in Spain. This coincided with what was then a major article of faith in the Community, that of 'auto-sufficiency', which was a driving force behind the Common Agricultural Policy (CAP) and therefore also of the burgeoning CFP. What lay behind the concept of 'auto-sufficiency' was the fact that the people of mainland Europe had starved after the end of the 1939–45 war; the CAP was to make sure that, whatever happened elsewhere, the inhabitants of the Community would never starve again. No matter if the Community produced too much food for its own use, the rest of the world would be only too eager to buy its surpluses because, as all the experts were forecasting at that time, there was going to be a world shortage of food. It is too easy to forget the factors which influenced the development of a policy when the circumstances have changed and are no longer relevant to those of today.

This emphasis on catching more fish can be seen in Article 11 of Regulation No. 2908/83 which provided that the purchase or construction of new vessels, the replacement of vessels over 12 years old and those which had sunk as well as those based in coastal areas where fishing is traditionally an important economic activity 'shall enjoy priority in consideration for Fund aid'. In hindsight it may be questioned why the almost certain outcome, overfishing of the stocks, of such a policy was not foreseen. The answer is very simple. Prior to the establishment of DG XIV, there was no scientific expertise in fisheries in the Commission, nor was there any until 1978. For eight years there was no-one to warn that adding more fishing effort to the already heavily over-exploited stocks in Community waters would produce less fish in the long term, not more. Also, there was no conservation policy so there was no need to consider the implications of the structural policy for another policy which did not even exist. Even when fisheries scientists were recruited to the staff of DG XIV and warned of the dangers of what was happening, they were ignored. In submitting an application for a FEOGA grant a fisherman had to prove that a project would be financially viable but the only proof needed was estimates of income and expenditure to show that a profit would be made with the new or modernized vessel. The Commission made no evaluation of the effect of the increased fishing capacity on the fish stocks and therefore on future earnings. Besides, nearly all the Member States were benefiting from the structural policy and did not want the flow of funds stopped. Only Italy and, later, Greece felt that they were being discriminated against because aid for construction and modernization was limited to vessels between nine and 33 metres long between perpendiculars, which excluded much of their fleets. This was purely an administrative convenience to limit the number of applications to a level with which the small number of Commission staff could process. Italy and Greece did not want the flow of funds stopped, only the tap opened wider for them, which they achieved by a modification to Regulation No. 4028/86 adopted in 1990 which reduced the lower limit to five metres.

The result was as might be expected, the size of the Community fleet increasing enormously from 1970 onwards. In 1970, when the structural policy was adopted, the estimated gross registered tonnage (GRT) of the Community fleet was 794 000 GRT; by 1983 it had increased by 64% to 1 303 000 GRT and by 1987 to 1 618 519 GRT, more than twice its size in 1970. In terms of engine power, which is a better index of fishing capacity than tonnage, the increase was even greater. In 1970 the estimated engine power was 1 983 000 kilowatts (kW); by 1977 it had increased to 3 691 000 kW and by 1987 to 6 476 471 kW, a more than threefold increase since 1970. The data prior to 1987 are very unreliable because there was no consistent method of collecting and recording the data until very recently and, in the case of tonnage, data were not reported between 1970 and 1987. Despite this reservation, the data show that there was an enormous increase in the size of the Community's fleet prior to the finalization of the CFP and subsequently, much of which would presumably not have occurred if Community finance had not been readily available. The data probably underestimate the increase in fishing pressure on stocks in Community waters for two reasons. First, the figures for 1987 do not include any data for Italy or the tonnage of the Dutch fleet. Second, and much more importantly, prior to 1977 many Community vessels fished in what were to become the waters of third countries. From 1977 onwards these vessels formed a decreasing proportion of the Community's fleet. What is not open to question is that the rates of fishing on the most important stocks increased considerably (Table 2.1), only decreasing on those stocks of herring which had collapsed and which were, for this reason, subject to severe restrictive measures. For the stocks of demersal species the average increase was 19%, excluding those of whiting in ICES divisions VIIa and VIId (see the Frontispiece for a map of the ICES area and its divisions).

However, as far as the Community and the fishermen were concerned, the policy was proving very successful. Not only were they able to purchase new vessels but their catches from many stocks, although not all, were increasing (Table 2.2). However, these increases in catches did not represent a failure of one of the fundamental laws of fisheries management, which is that increasing the amount of fishing on stocks which are already over-exploited results in catches falling in the long term (see Chapter 8). These increases resulted from much larger numbers of young fish than normal coming into the fisheries for many stocks, especially those for North Sea and west of Scotland cod,

Table 2.1 Fishing mortality rates, expressed as percentages, on the most important stocks in Community waters in 1973 and 1983.

Stock	1973	1983	% change
Herring IV	68	27	− 60
Herring VIa (North)	45	38	− 16
Herring VIa (South)	18	31	72
Herring, Clyde	38	25	− 34
Herring VIIa	39	16	− 59
Herring (Celtic Sea)	43	41	5
Mackerel (Western stock)	4	19	375
Cod IV	50	59	18
Cod VIa	44	55	25
Cod VIIa	55	57	4
Cod VIIf,g	41	49	20
Haddock IV	54	62	15
Haddock VIa	48	41	− 15
Whiting IV	48	51	6
Whiting VIa	73	41	− 44
Whiting VIIa	56*	66	(18)
Whiting VIId	27†	59	(119)
Saithe IIIa, IV	28	46	64
Saithe VI	30	27	− 10
Plaice IV	31	36	16
Plaice VIIa	45	42	− 7
Sole IV	36	38	6
Sole VIIa	29	29	0
Sole VIIf,g	18	30	67
Sole VIId	22	37	68
Sole VIIe	18	36	100

Source: Report of ACFM for 1990; rates of fishing mortality for years other than 1983 extrapolated from graphs.
* data for 1980; † data for 1977 (data for earlier years not available).

haddock and whiting, which started in 1963, a sequence of events described in Chapter 7. The boom burst in the mid-1980s causing the manner in which the structural policy was implemented to change direction violently. The key theme of the policy then became multi-annual guidance programmes.

Multi-annual Guidance Programmes

Provision for multi-annual guidance programmes (MAGPs) was made in Regulations No. 2141/70 and No.101/76 but only briefly, in one indent of one article. They were seen as part of a means of co-ordinating structural policies for the fishing industry, enabling financial assistance to be better planned. By 1983, MAGPs had become a major feature of the structural policy. This is

Table 2.2 Nominal catches by Member States, in thousands of tonnes, from selected stocks, 1973 and 1983.

Stock	1973	1983
Mackerel (Western stock)	77	428
Cod IV	216	230
Cod VIa	12	21
Cod VIIa	11.8	10.3
Cod VIId,e	2.0	4.4
Cod VIIf,g	3.0	5.3
Haddock IV	133	159
Haddock VIa	28	31
Hake (Northern Stock)	30	28
Whiting IV	132	99
Whiting VIa	5.1	16.7
Whiting VIIa	10.3	10.8
Whiting VIId, e	5.6	7.4
Whiting VIIf,g	7.0	8.5
Saithe IV	103	76
Saithe VI	33	26
Plaice IV	130	103
Plaice VIIa	5.1	3.7
Plaice VIId, e	3.1	6.2
Plaice VIIf, g	1.0	1.2
Sole IV	19.3	20.0
Sole VIIa	1.4	1.3
Sole VIIf, g	1.4	1.4
Sole VIId, e	1.5	4.5

Source: Reports of ACFM.

reflected by the fact that MAGPs became a title, a major section of any regulation, comprising five articles of Regulation No. 2908/83. Their objective was clearly defined as achieving:

'in respect of the fishing sector, a satisfactory balance between the fishing capacity to be deployed by the production facilities covered by the programmes and the stocks which are expected to be available during the period of validity of the programme'.

The information which had to be included in the MAGPs, which also covered aquaculture, was also specified in detail.

A major problem and one which has yet to be satisfactorily resolved is how to measure fishing capacity. The Commission chose to use two units, gross registered tonnage (GRT) and engine power, measured in kilowatts (kW). Neither is satisfactory for several reasons. Although both units are measures of the size of the fleets, they are only indices of their potential to catch fish. They are not directly related to how much fish the fleets can catch, which depends upon the amount of time the vessels spend fishing and often the quantity of gear which they use. For this reason there is no fixed relationship between either unit and fishing mortality rates, the parameter by which fishery scientists measure the rate of fishing on the stocks. Nor is there any mathematical model by which this can be done. Gross registered tonnage also suffers from the drawback that the rules which specify how it is to be measured can result in boats with identical external hull design having different gross registered tonnages, depending upon their internal construction; the rules exclude certain internal spaces. However, this problem will be resolved when the Community changes to using gross tonnage in 1994, under which this possibility disappears. Engine power also suffers from the disadvantage that installed engine power is difficult to measure and, as experience in enforcing regulations based upon engine power has shown, easy to manipulate. However, despite their disadvantages, both units are internationally defined with standardized rules for their measurement. There can be no dispute between Member States for this reason. Even though they may not be totally satisfactory, they provide a measure of fishing capacity which is common to all Member States. In the final analysis, a vessel which does not exist cannot fish.

Table 2.3 Fleet size objectives established by the multi-annual guidance programmes for the period 1983–86 compared with the situation as at 1 January 1987.

Member State	Unit	Objective 31 Dec 1986 (a)	Situation as at 1 January 1987 (b)
Belgium	GRT	22 000	25 165
	kW	70 656	78 506
Denmark	GRT	122 879	136 680
	kW	525 241	563 667
Germany	GRT	78 479	51 500
	kW	161 494	139 100
France	GRT	207 560	207 560
	kW	1 158 576	1 158 576
Greece	GRT	134 659	137 761
	kW	502 467	568 823
Italy	GRT	275 255	302 986
	kW	1 568 288	1 796 829
Ireland	GRT	45 300	58 845
	kW	181 200	234 892
Netherlands	GRT	66 800	82 400
	kW	390 080	498 800
UK	GRT	146 000	163 410
	kW	763 515	840 982

GRT = gross registered tonnage; kW = kilowatts (engine power).
Sources: (a) Commissioner Decisions 90/101–6/EEC and 90/229/EEC (for France); OJEC L66 of 14.3.1990 and L124 of 15.5.1990; (b) Commission document SEC (90) 2244 final.

MAGPs for the period 1982–86 were drawn up under this programme but all Member States except France and Germany failed to meet their targets. (Table 2.3). There were many reasons for this. Being the first programmes, neither the Commission nor the Member States had any previous experience of implementing such programmes. But even more importantly, the necessary records of fleet composition in every Member State were incomplete. Furthermore, in 1982 the prevailing attitude was that there was no urgent need, if any, to reduce the size of the fleets. By the time that the second set of MAGPs came to be formulated, for the period 1987–91, it had become obvious that all was far from well with the majority of fish stocks. TACs were having to be reduced and in particular the state of the North Sea and west of Scotland stocks of cod and haddock were causing serious concern with the Advisory Committee on Fishery Management (ACFM) of the International Council for the Exploration of the Sea (ICES) issuing warnings of ever increasing seriousness and urgency. The Commission therefore made a serious effort to agree MAGPs which would stop the ever increasing expansion of the size of the Community's fleets.

In order to have a basis for setting the targets, the Commission convened an independent group of experts who reported in 1990. In the absence of any definition of what the Commission meant by 'fishing capacity in balance with resources', the experts took as their yardstick the concept of F_{max}, the fishing mortality rate at which the maximum long-term sustainable yield is obtained under the existing pattern of fishing (see Chapter 8). Although the situation varied from area to area, the group concluded that the majority of fish stocks were over-exploited and that a reduction in fishing capacity – in fact, in its terms, of fishing mortality rates – of not less than 40% on average was needed to meet the objective of fishing capacity in balance with resources. For some stocks, such as North Sea cod and haddock, the required reduction was 70%. The group advised that fishing capacity should be brought into balance with the fishing mortality rates which correspond to those on which present TACs are fixed by the end of 1996 at the latest and preferably by the end of 1994. The experts provided no guidance on how fishing capacity and fishing mortality rates were to be equated.

In determining the programmes for the period 1987–91, the Commission made no attempt to achieve large reductions in fishing capacity. Instead, the Commission took a very realistic view of what was politically possible, agreeing reductions of only 3% in tonnage and 2% in engine power on average. The Commission recognized that increases in efficiency would mean that these reductions, even if met, would result in an actual increase in fishing capacity. But it regarded this as a first step in accustoming the Member States to the more drastic measures which would need to be implemented from 1987 to the end of 2001.

By 31 December 1991 only five Member States (Denmark, Germany, Spain, Italy and Portugal) had met their objectives in respect of both tonnage and engine power, while another two (France and Ireland) had nearly met their targets (Table 2.4). However, four Member States, Belgium, Greece, the Netherlands and the UK had failed to meet their objectives by margins of 11%

or more. The fleet sizes of Belgium, Greece, the Netherlands and the UK on 31 December 1991 were greater in both terms of tonnage and kilowatts than their objectives for 31 December 1986, although the UK had managed to have its objectives revised upwards in its programme for 1987–91. This situation also applied to Germany but the data for 1986 apply to the former Federal Republic of Germany and the later data to the reunified Germany. The fleet size of Ireland was also greater at the end of 1991 than the end of 1986 but in terms of tonnage only. However, even for those Member States which met their targets, it must be a matter of some doubt as to what extent active fishing capacity had been reduced; it seems more than likely that many Member States met their targets by weeding out from their registries vessels which were fishing very little, if at all.

Table 2.4. Objectives for multi-annual programmes compared with actual situation.

Member State	Unit	Objective 31 Dec 86	Objective 31 Dec 91	Situation 31 Dec 91	Objective 31 Dec 96
Source		(a)	(b)	(b)	(b)
Belgium	GRT	22 000	21 340	27 089	17 992
	kW	70 656	69 242	79 816	58 512
Denmark	GRT	122 879	119 188	114 926	108 422
	kW	525 241	514 716	488 278	435 738
Germany	GRT	78 479	85 336	78 341	74 764
(c)	kW	161 494	206 465	190 273	183 856
Greece	GRT	134 659	130 946	162 395	123 014
	kW	502 467	493 776	710 899	471 532
Spain	GRT	—	673 303	644 989	618 773
	kW	—	1 955 372	1 910 145	1 810 836
France	GRT	207 560	201 604	195 969	180 378
	kW	1 158 576	1 055 050	1 072 428	949 087
Italy	GRT	275 255	268 198	267 471	249 182
	kW	1 568 288	1 541 664	1 536 518	1 464 680
Ireland	GRT	45 300	48 750	50 693	51 195
	kW	181 200	197 011	176 075	179 732
Netherlands	GRT	66 800	(d)	(d)	(d)
	kW	390 080	382 878	441 953	346 888
Portugal	GRT	—	186 449	165 447	167 503
(mainland)	kW	—	461 143	433 549	428 061
UK	GRT	146 000	193 027	214 733	173 455
	kW	763 515	1 095 206	1 228 922	995 627

Sources:
(a) As Table 2.3.
(b) Commission decisions on multi-annual programmes for the fishing fleets of the Member States pursuant to Council Regulation 4028/86 (92/588–598/EEC) OJEC L401, 31.12.1992, pp 1–70.
(c) Data for 1992–96 are for reunified Germany.
(d) Targets set in kilowatts only.

Faced with this situation and the serious over-exploitation of most of the stocks of demersal species, the Commission attempted to agree very large reductions in fleet capacity for the period 1992–96. It was widely reported that the Commission was seeking reductions of the order of 30%. France bitterly opposed such a large reduction. Ireland went even further, threatening to initiate a case against the Commission in the Court of Justice under Article 173 of the Treaty which relates, amongst other issues, to the misuse by an institution of its powers. Amongst the arguments put forward by Ireland was that it is unable completely to utilize its quotas with its existing fleet and that the proposals of the Commission would have resulted in a loss of 1000 jobs, 12% of the country's fishermen. Faced with this and similar opposition from other Member States, the Commission was unable to agree programmes for the whole of the period 1992–96 by the end of 1991. While negotiations continued, a set of programmes for 1992 only were adopted. Programmes for the period ending 31 December 1996 were agreed in December 1992; they are summarized in Table 2.4. These programmes are much more detailed than those which preceded them, different objectives being set for each sector of each national fleet. These sectoral objectives are a 20% reduction in fishing capacity for those fleets using bottom trawls to fish for demersal species, a 15% reduction for those using bottom trawls and dredges to fish for benthic stocks, such as *Nephrops*, and no reduction for those fishing all other stocks. These reductions are much less than the Commission sought but doubtless it had sought decreases which were far larger than it expected to agree in order to achieve significant cutbacks in fishing capacity. Another feature of the programmes is that 25% of the required reduction has to be made annually.

Given the fact that over half the Member States failed to achieve the small reductions in fishing capacity set by the programmes for 1987–91, there must be serious doubts about whether the much bigger reductions set by the programmes for 1993–96 will be achieved. There will be a major problem in monitoring what happens because the decisions allow up to 45% of the reduction in capacity to be achieved by measures other than reductions in the number of vessels, such as limitations on days at sea and increases in minimum mesh sizes. The latter only reduce the rate of fishing on the youngest age groups, whereas the objective is, or should be, to reduce the rate of fishing on all age groups (see Chapter 6).

All the programmes include provisions for review. This is particularly relevant to Ireland because the Commission is committed to examine its claim that its fleet should be allowed to increase even further. The programme for Ireland permits an increase from the situation at the end of December 1991 (Table 2.4), the only Member State allowed an increase in its fishing capacity. As a result, Ireland did not proceed with its case before the Court of Justice. One of the arguments used by Ireland, that of under-utilization of its quotas, arises from the political decision made when agreeing the conservation policy in 1983 to fix TACs much higher than recommended by the Advisory Committee on Fishery Management in order to respect the Hague Preferences of Ireland, and also the UK. This is described in detail in Chapter 3. Although the TACs were fixed at more realistic levels from 1991 onwards, the aftermath of this decision

Table 2.5 Number of projects financed under Regulation No. 4028/86 during the period 1988–92.

Year	Construction	Modernization	Aquaculture	Artificial constructions	Total
		Type of project			
1988	28	697	266	1	992
1989	227	594	177	2	1000
1990	105	482	256	2	845
1991	28	591	226	8	853
1992	54	624	188	6	872

Table 2.6 Total number of projects (upper line) and number of construction projects (lower line) financed by Member States during the period 1988–92.

Member State	Year				
	1988	1989	1990	1991	1992
Belgium	8	8	12	5	9
	0	1	1	0	0
Denmark	117	105	51	26	28
	0	0	0	0	0
Germany	15	35	47	31	86
	0	11	7	3	8
Greece	42	60	59	66	86
	0	10	3	0	0
Spain	348	309	329	302	250
	0	82	48	11	32
France	79	89	77	82	123
	14	27	11	0	6
Ireland	46	52	56	54	46
	0	8	8	0	0
Italy	133	108	70	153	121
	0	57	14	12	5
Netherlands	15	23	21	8	1
	0	0	0	0	1
Portugal	56	86	65	66	46
	14	28	10	2	3
UK	133	125	58	60	76
	0	3	3	0	0
Totals	992	1000	845	853	872
	28	227	105	28	54
ECU million	67.9	115.8	93.7	68.7	59.5

Source: Official Journals of the European Communities.

Table 2.7 Expenditure on structural measures, 1973–1990 (ECU millions).

Year	1973	1974	1975	1976	1977	1978	1979	1980	1981
Construction	—	—	—	—	—	2.90	11.63	13.51	25.42 (1)
Modernisation	—	—	—	—	—	—	—	7.86	10.37
Adjustment of capacity	—	—	—	—	—	—	—	—	—
Other	—	6.85	6.85	6.85	6.85	26.20	31.20	25.70	49.80
Sub-total	0.00	6.85	6.85	6.85	6.85	29.10	42.83	47.07	60.17
Aquaculture	7.14	16.27	13.76	18.57	6.16	2.02	3.34	6.42	(1)
Total	7.14	23.12	20.61	25.42	13.01	31.12	46.17	53.49	60.17

Year	1983	1984	1985	1986	1987	1988	1989	1990
Construction	21.30	39.10	46.80	38.70	64.00	8.70	63.50	44.20
Modernisation	7.00	10.40	15.20	18.00	9.20	19.30	20.30	26.20
Adjustment of capacity	6.85	6.85	6.85	6.85	26.20	31.20	25.70	49.80
Other	—	—	2.30	3.90	1.90	16.80	14.60	13.00
Sub-total	35.15	56.35	71.15	67.45	101.30	76.00	124.10	133.20
Aquaculture	4.80	10.30	8.50	20.40	24.20	40.00	37.70	47.50
Port structures	13.60	11.80	24.70	30.10	25.80	27.20	48.70	52.00
Total	53.55	78.45	104.35	117.95	151.30	143.20	210.50	232.70

Notes:

(1) Construction and aquaculture combined: expenditure not available separately.

(2) 1971–79 EEC 9; 1980–85 EEC 10; 1986–92 EEC 12

Sources:

1973–81 Booss, D., European Fisheries. Situation and outlook after the adoption of the common fisheries policy. *Diesel Engine Journal.* No. 53, 23–28.

1983–90 Commission document SEC(91) 2288 final.

1991–92 Official Journals of the European Communities.

still exists. On a scientific basis the stocks are over-fished and the level of exploitation needs to be reduced yet, politically, Ireland has a case for increasing the size of its fleet. The situation is aggravated by a report on the state of the Irish fisheries compiled by external consultants employed by the Commission; the report argues that the Irish fisheries should be developed. The Commission is in a difficult position because, if it agrees that the Irish fleet should expand, those of other Member States, notably the UK, would have to decrease if the stocks were not to become further overexploited.

One of the difficulties which the Commission faces in implementing these programmes is that, although they are legally binding on the Member States, being Decisions of the Commission (see Chapter 1), the only real means by which the Commission can enforce them is not to approve applications for grants for new construction from those Member States which do not meet their programmes. The Commission has been rigorous in applying this policy between 1988 and 1992, as shown by the number of projects granted for construction (Table 2.5). For the most part, only those Member States which have met their targets under their MAGPs have had construction projects approved (Table 2.6). The total number of projects approved has also declined, to less than 25% of its 1989 level. Table 2.7 provides a longer term view of expenditure, not only on grants for the construction of new vessels but on the structural policy as a whole, showing how the emphasis has changed with time. In particular, it illustrates how the Community was prepared to continue spending a very high proportion of the structural budget on vessel construction until 1990 even though every indicator clearly signalled that more fishing vessels were the last thing that the Community needed.

The data in Table 2.7 need to be treated with some caution because expenditure on structural measures is not published in the Official Journals of the European Communities and the data which are available, even in Commission documents, are not always consistent. Furthermore, the headings under which expenditure is listed may also not always be consistent. (The same comments apply to Table 2.8.) However, they are the only data available which show long-term trends in expenditure.

Although programmes have been agreed for 1993–96, it is clear the battle between the Member States and the Commission on this issue is far from over. There still promises to be a continuing major political fight on how the structural policy is to develop. Even if sufficient Community funds are made available for the amount of decommissioning required to meet the objectives, which now seems very unlikely given the decisions at the European Summit in Edinburgh, the social consequences will be so severe that Member States are unlikely to accept further significant reductions, if any. They will not be slow to point out that the regulation states that the programmes must 'take account of the socioeconomic consequences and the regional impact of the developments foreseen in the sector concerned' and that Regulation No. 3760/92 provides that the conservation policy shall be implemented 'in appropriate economic and social conditions'.

What of the other objectives of the policy? Aquaculture is still actively encouraged (Tables 2.5 and 2.7) both because it is seen as helping to reduce the

Community's deficit in fish supplies and because much of the budget is spent in the southern Member States which have not benefited as much as the northern Member States from grants for building and modernization of vessels. Many optimistically believe that supplies from aquaculture could one day replace declining resources from the sea. This ignores the fact that most of the finfish species used in aquaculture (salmon and seatrout) are predatory species. An essential ingredient of their food is protein of fish origin. This fish protein originates partly from processed fish offal but mainly from the industrial fisheries which catch species for reduction to meal and oil. Put in other terms, most finfish reared by aquaculture represent no more than reprocessed fish taken from the sea in the industrial fisheries. If the pressure to reduce or ban totally the industrial fisheries in the Community was successful, there would be a negative impact on aquaculture. Supplies could be imported but presumably costs would rise.

In order to reduce fishing capacity in Community waters, much emphasis is being placed upon exploratory fishing voyages, redeployment operations and, in particular, joint ventures and joint enterprises. However, exploratory voyages look to have little long-term potential. New stocks which provide economically viable fisheries are likely to be few and far between and, if they are found in international waters, past history shows that they will be rapidly depleted by fleets from every nation. Joint ventures may prove far more effective in removing fishing capacity.

There are few new or underexploited stocks available to Community fishermen. The stock of blue whiting in the northern North Atlantic is underexploited and could probably sustain an annual catch of about 800 000 t a year compared with total catches of about 500 000 t from 1988 to 1992. The main reason for which the stock is underexploited is that it has not proved possible to process the species for human consumption. Also, blue whiting has a low oil content so that it is not economical to catch for reduction to meal and oil. Additionally, the stock is not easily accessible, being in Community waters and then up to 200 miles west of Ireland in the first quarter of the year when weather conditions are at their worst.

There is increasing interest in stocks of fish in the waters to the west of the British Isles deeper than 1500 m. These have been fished by French vessels for a decade. However, these stocks cannot absorb much fishing. The present attractiveness of the fishery is based upon high catch rates resulting from exploiting virgin stocks. Once the accumulated surpluses have been caught, catch rates will fall. It is known that the species concerned are slow growing and, in consequence, the stocks will be replaced only very slowly and catch rates will remain low indefinitely. The 'search for new outlets for surplus for products derived from surplus or underfished species' is unlikely to make little contribution to the problem of over-capacity.

An evaluation

The structural policy has swung from one extreme to another. It began as a policy which was regarded as having little or nothing to do with fish stock conservation. Now the policy is seen as an integral part of the conservation policy, as explicitly recognized for the first time in Article 1 of Regulation No. 3760/92. The policy's initial, overriding objective was vessel construction; now it is vessel destruction.

Originally the policy provoked little or no political controversy because all Member States benefited from the grants and subsidies available, even if Italy and Greece felt that they were discriminated against. From its start in 1970 to the mid-1980s the policy was considered a political success. While grants were freely available, fishermen were very pleased with the operation of the policy and therefore placed no pressure on ministers, the usual measure of 'political success'.

However, the drastic decline in catches from many of the Community's most important fish stocks forced a reappraisal of the policy. For 1991 and 1992 ACFM recommended that fishing effort on the stocks of North Sea and west of Scotland cod, haddock and whiting should be reduced by 30%. This was the first time that ACFM had expressed its advice in terms of reducing fishing effort rather than rates of fishing mortality. The Commission, at least, belatedly realized that the implementation of the structural policy was crucial to the conservation policy.

However, the damage of the years 1970 to 1990 has been done and the modern vessels which were built with the help of Community aid exist. The question is how can this be rectified? The means which the Community has adopted are MAGPs and decommissioning. However, the MAGPs for the period ending in 1991 were only partially successful and have done little more than stabilize fishing capacity at its 1987 level, if that. The real test of the policy will come from 1993 onwards. Will the Member States meet their objectives to which they have agreed in the MAGPs for 1993–96? Will the Community budget available for decommissioning be sufficient to pay for the number of vessels which need to be withdrawn? Herein lies a major weakness of the structural policy; there has never been an evaluation of the amount of money which is required to induce a fisherman to decommission his vessel. Faced with the choice between a decommissioning grant which is too small to make his future livelihood certain and continuing to fish, many may choose to do the latter, unless forced out by other means.

Now that grants are no longer freely available for vessel construction, if at all for fishermen of those Member States which have not met the objectives set by their MAGPs, fishermen are complaining and political pressure is beginning to mount. A further cause for dissatisfaction is that the money available for decommissioning is too little. Not only the fishermen are dissatisfied; the ship building industry is also dissatisfied because orders for new vessels have fallen drastically and yards have become bankrupt. There are some who think that building fishing vessels is an industry in itself, even if there are no fish for the vessels to catch. Given all the circumstances it is inevitable that the political temperature will continue to rise.

Although the policy has been very successful in enabling the Community to build a very modern fleet, its very success in this respect has also been its own failure because of the consequences, not only for the fish stocks but for the economics of the industry, an issue discussed in detail in Chapter 12.

The markets policy

The policy

Regulation No. 2142/70 provided for a system of market support which was almost identical to that existing for agricultural products, not surprisingly in view of the 'agricultural' background to fisheries. This regulation was replaced at the end of 1992 by Regulation No. 3759/92 whose basic provisions are effectively unchanged. Although all the regulations have been very detailed and have become increasingly complex with each succeeding version, their fundamental elements have been very simple.

The main objectives of the policy are:

• to establish marketing standards;
• to stabilize market prices and avoid the formation of surpluses;
• to help support producers' incomes; and
• to consider consumers interests

by providing for:

• the specification of marketing standards by freshness and size categories for the main commercial species;
• the fixing of guide prices; and
• the management of marketing by producers' organizations.

The guide prices are based upon the average prices at first-hand sale in the Community over the last three years. The producers' organizations, for whose establishment Community funds are available, are authorized to fix withdrawal prices which lie within the range of 70–90% of the guide price, below which fish cannot be sold but must either be withdrawn from the market, either to be destroyed or, in the case of high value species, to be held in cold storage, for which 'carry-over aid' is available. The Community reimburses the withdrawal costs. Based upon the guide prices, reference prices are fixed below which imports became subject to duty in order to prevent dumping of fish on the Community markets.

An evaluation

The objectives of the markets policies have been largely achieved, almost entirely due to the fact that the Community has an increasing deficit in supplies of most fish species, the most notable exceptions being anchovy and sardine. This deficit has grown in tonnage from 703 thousand tonnes in 1984 to 1709 thousand tonnes in 1990 and in value from ECU 2.28 billion to ECU 5.23 billion over the same period. In consequence, there have very rarely been occasions on which fishermen could not sell their fish at high prices, except during periods of temporary oversupply, such as notably the beginning of 1993.

Unlike the system of prices in the Common Agricultural Policy, this policy has always been implemented to discourage fishing for withdrawal, the equivalent of growing crops for intervention. A major step in this direction was the adoption of provisions laying down a sliding scale for financial compensation which is 85% of the withdrawal price for the initial 5% of quantities put up for sale by the producers' organizations, 70% for quantities between 5% and 10% and nothing for those exceeding 10%. Also, from the outset, withdrawal prices have also been fixed low in order to discourage fishing for intervention. But it is unlikely that this could have been achieved if there had been a glut of supplies.

A measure of the success of the policy is that its costs have remained almost stable from 1980 onwards, with annual expenditure significantly exceeding ECU 25 million only in one year, 1978 (Table 2.8). Between 1983 and 1990 the expenditure on the markets policy represented only 10% of the total budget for fisheries. Politically, the policy has very rarely created tension, the annual prices often being agreed in COREPER. This is in total contrast with the situation with the annual fixing of farm prices, which has been often accompanied by the presence of thousands of farmers milling around the Charlemagne building where the Council meets, blowing whistles and letting off fireworks, with riot police on the streets and the smell of tear gas drifting through the air.

The external fisheries policy

The policy

This policy did not come into effect until 1977 and is discussed here only for convenience. When it became obvious that one of the most probable outcomes of the third United Nations Conference on the Law of the Sea would be the recognition of the right of coastal states to establish 200-mile fishery limits, the Commission recognized that an essential part of the CFP would be a policy which covered external fisheries relations. The policy came into being with the general extension of fishery limits to 200 miles on or about, depending upon country, 1 January 1977. The Commission had two objectives in including international fisheries relations within the CFP. One objective was, by making the Community responsible for international fisheries agreements, to prevent the Member States from individually negotiating fisheries agreements with third countries in which it was feared that they would be played off against each other; the other objective was to ensure that as many distant-water vessels as possible were able to continue to fish in the waters of third countries in order:

(1) to maintain supplies of fish caught by Community vessels, as opposed to imports, the Community having a deficit in fish supplies;

(2) to keep Community fishing vessels fishing in what had been international waters so that they should not return to Community waters, thus placing further pressure on the already over-fished stocks in Community waters;

(3) to maintain supplies of species which do not occur in Community waters or for which the Community has a deficit; and

Table 2.8 Expenditure on markets policy, 1973–1990 (ECU millions).

Year	1973	1974	1975	1976	1977	1978	1979	1980	1981
Withdrawals	0.62	0.51	6.50	6.71	4.61	6.85	8.50	11.56	15.39
Export subsidies	0.57	0.66	2.79	3.76	3.30	7.17	8.51	11.40	12.63
Producers' organizations		0.01	0.14	0.14	0.18	0.09	0.20	0.08	
Total	1.19	1.18	9.43	10.61	8.09	14.11	17.21	23.04	28.02

Year	1982	1983	1984	1985	1986	1987	1988	1989	1990
Withdrawals	ND	17.15	14.54	18.41	17.23	17.38	47.05	23.93	23.60
Export subsidies	ND	8.30							
Producers' organizations	ND		0.05	0.08		0.07	0.21	0.03	0.01
Total	ND	25.45	14.59	18.49	17.23	17.45	47.26	23.96	23.61

Sources:
As Table 2.7.
ND: no information available.

(4) to minimize unemployment in the distant-water sector.

At the same time as the Council made a decision that all Member States would extend their fisheries limits to 200 miles as from 1 January 1977, with the exception of the Mediterranean, the Baltic Sea and the Skagerrak and Kattegat, it also adopted a decision authorizing the Commission to negotiate fisheries agreements with third countries. The policy is therefore not covered by a basic regulation, agreements made to implement this policy being based upon Article 43 of the Treaty, which concerns the general provisions for implementing the CFP, or Article 113 if they concern trade.

To date, the initiatives for implementing the policy have come mainly from those Member States which wish to continue their traditional fishing activities in third country waters. The manner in which the policy is implemented is that the Council adopts a decision authorizing the Commission to negotiate an agreement on behalf of the Community. Once the Commission has negotiated what it considers is a satisfactory agreement it presents a proposal embodying the agreement to the Council which either adopts it or orders the Commission to re-negotiate it. Representatives from the Member States attend the negotiation meetings so, in general, the agreements are usually adopted with very little debate.

The policy is very important for the majority of Member States, although their interests are different. The agreements which permit fishing in the waters of, in the main, African countries, are most important to Greece, Italy, Spain and Portugal and those which allow fishing in the waters of the Faroe Islands, Greenland, Norway and Sweden are mainly of interest to Germany and the UK. France is interested in both.

This policy is almost independent of the other three policies which comprise the CFP except in so far as it concerns the agreements with Norway. Many of the fish stocks in the North Sea inhabit both the Community and Norwegian fishery zones; they are so-called 'joint stocks' whose management is shared between the two parties. In particular, the TACs and the share of these TACs have to be agreed annually. The means by which these agreements are reached is that the Commission negotiates on behalf of the Community, presenting a proposal embodying the agreement to the Council for adoption. On several occasions during the negotiation of the conservation policy, described in Chapter 3, the Council refused to adopt these proposals because to have done so would have fixed the TACs for several important stocks which were the subject of the general allocation debate. This was more of a negotiating ploy, not a question of principle concerning the agreements themselves.

An evaluation

The policy has been successful to date, agreements with 27 third countries, of which 19 are with African states, having been concluded. This has enabled the objectives of the policy to be met, in particular that of maintaining employment in the distant-water sector, which would have given serious political problems otherwise. Whether, if it had not succeeded, many of the vessels fishing distant waters could have been economically redeployed in Community waters is, however, open to question. The policy has been very expensive, costing on average ECU 68 million (£48 million) a year in the period 1983–90, which represents 29% of the fisheries budget, with 41% of the budget being taken by the agreement with Morocco and 20% with Greenland[1]. There are serious doubts as to whether the value of the fish caught exceeds the costs of the agreement.

That part of the policy which concerns Norway and Sweden is likely to become of increasingly less importance as these countries accede to the Community and the problems which they now pose are transferred to the internal policy.

The major external issue, which the Community does not yet appear even to have recognized, is likely to be the need to extend fishery limits. This issue will arise from the increasing amount of fishing by non-Community vessels in waters outside the Community's 200-mile fishery limits. The position of the Community has always been that in international law, the management of the fish stocks occurring outside 200-mile fishery limits should be regulated by international fishery commissions. This position was reinforced by the dispute which the Community had with Canada over the management of the northern cod stock. This is a transboundary stock, most inhabiting Canadian waters, the remainder occurring outside the Canadian 200-mile limit on the Grand Bank. Because most of the stock occurs in Canadian waters, Canada claimed an overriding interest in the stock. Under what the Commission claimed to be Canadian pressure the Northwest Atlantic Fisheries Organization (NAFO) fixed a zero TAC for that part of the stock outside the Canadian 200-mile fishery limit, effectively giving Canada the unilateral right to fix the TAC for the stock. The Community objected to the recommendation to which, under the rules of NAFO, it was not bound, thus allowing Community vessels to continue to fish to quotas unilaterally fixed by the Council. This dispute was resolved by the Community accepting in 1992 the recommendation of NAFO for a zero TAC.

However, the Community is increasingly experiencing the same problems position as Canada faced with the Community. A growing number of non-Community vessels is fishing in the waters to the west of the British Isles outside 200-mile limits on stocks which are transboundary stocks. This will progressively increase the exploitation of these stocks and render Community conservation measures useless. The situation is made more complex by the fact that the UK claims a 200-mile fishery limit around Rockall for which there is no basis under the 1982 UN Convention on the Law of the Sea because Rockall is uninhabited. It is improbable that the UK would win a case brought before the International Court of the Hague on this issue. Thus, the Community may well find itself in a position in which its interests are to initiate moves to extend fishery limits beyond 200 miles, maybe as far as the edge of the Continental Shelf. This issue promises to generate much heat in the future and its solution will have global political implications.

Note

(1) The rate of exchange used throughout this book is that prevailing prior to the UK leaving the Exchange Rate Mechanism, which was ECU 1 = £0.7.

Chapter 3

The Building of a Conservation Policy

or The Marathon Negotiation

Introduction

The political background

The original six Member States had little interest in fish stock conservation. As already described, to the extent that they had an interest in fisheries it was primarily in establishing structural and market policies and secondarily in ensuring that access to their traditional fishing grounds was maintained if the international regime concerning fisheries limits changed. The Commission's paper of 1967 had recognized the need for a conservation policy, without which an integrated Common Fisheries Policy could not be effective, ideas which had been embodied in Regulation No. 2141/70, but nothing happened until the accession of Denmark, Ireland and the UK gave the necessary impetus. With the accession of the UK, the Community had, for the first time, a Member State which had a long-established history of interest in fisheries conservation. In 1937 it had been one of the first countries to introduce minimum mesh size regulations and after the 1939–45 war had convened the London Over-fishing Conference which was the forerunner of the North-east Atlantic Fisheries Commission (NEAFC). The UK was, and still is, one of the leading countries in fisheries research and was instrumental in carrying out much of the research on which conservation legislation is based. Not only was the UK interested in introducing conservation measures but it also realized the need for their effective enforcement. The scene was set for establishing a conservation policy.

However, this policy did not commence with 'a clean sheet' but inevitably built upon the measures already adopted by NEAFC and more importantly, the provisions already adopted by the Community, however imprecisely these were drafted. As already noted, Regulation No. 2141/70 provided for 'access', TACs and technical conservation measures, all of which, together with control and enforcement, were to become the major issues to be settled under the conservation policy.

The starting gun to the negotiations on this policy was the emerging agreement at the UN Law of the Sea Conference that coastal states had the right to establish a 200-mile fishery. Prior to their establishment, there was no point in the Community adopting a conservation policy. To be effective, a conservation policy must apply to the whole area inhabited by the fish stocks to which it applies. There was, therefore, little point in the Community developing a conservation policy up to this point. In the Mediterranean no Member State has fishery limits greater than 12 miles, the main reason for which the conservation policy does not, as yet, apply there, although one is now being formulated. Whether as a result of sheer chance or the perspicacity of the Commission officials who had drafted Article 2 of Regulation No. 2141/70 on equal access, the existence of this article meant that, following the agreement of all Member States to extend their fisheries limits to 200 miles, except in the Mediterranean and Baltic Seas and in the Skagerrak and Kattegat, a Community sea to which a common conservation policy could be applied was automatically created as from 1 January 1977.

The proposal of the Commission

Although the Commission had set out in its paper of 1967 its ideas on what a conservation policy might include and Article 4 of Regulation No. 2141/70 made provision for certain conservation measures, these obviously reflected the general conservation measures in existence at that time. These measures, such as catch limitations and gear restrictions, were not specific to the needs of the Community, certainly not to a Community which now included three Member States with major fisheries interests.

In October 1976 the Commission presented its ideas on what a conservation policy should include by making a proposal for a Council Regulation establishing a Community system for the conservation and management of fishery resources. This proposal embodied all the elements which were eventually to form the final conservation policy. The objectives of this proposed regulation and the means by which they were to be achieved, as described in its preamble, were:

(1) the protection of fishing grounds in order 'that stocks are conserved and reconstituted';
(2) conservation measures 'which may involve limitations on fishing, rules for the use of resources, special provision for inshore fishing and structural measures';
(3) provision for total allowable catches (TACs) and quotas;
(4) special provisions for areas heavily dependent upon fishing (which were specified only as Ireland and the northern parts of the UK);
(5) quotas to be allocated between Member States on the basis of a reference period;
(6) the establishment of a system of control and enforcement;
(7) the establishment of a Scientific and Technical Committee for Fisheries.

Access

The background to the question of access has already been described in Chapter 2. The question was resolved for a period of ten years by the Act of Accession of Denmark, Norway, Ireland and the UK and was insignificant for the UK in the context of the political objective of acceding to the Community.

However, the UK inshore fishing industry, faced not only by the consequences of the accession of the UK to the Community but also economic decline caused by a combination of increasing fuel costs and falling income caused by imports, had organized itself and started a campaign for extended exclusive fisheries limits. This campaign reached its climax in 1975 with a series of blockades of ports. Amongst the promises made by the UK government which led to the blockade being lifted was one 'to seek changes in the Common Fisheries Policy of the EEC in the light of the United Nations Law of the Sea conference'. Up until this point, the CFP had been excluded from the renegotiation of the terms of accession of the UK to the Community to which the Labour government, elected in 1974, was committed.

The fishing industry initially argued for an exclusive fishery zone of 100 miles which was reduced to 50 miles. This formed the opening negotiating position of the UK which was eventually modified to 'dominant preference' within 50 miles. Ireland, whose fishermen had similar problems to those in the UK, aligned itself with the UK. The other Member States were totally opposed. Fishing plans and a licensing system within both of which dominant preference would be given to the coastal state were proposed by the Commission as compromise solutions. Neither were acceptable to the UK although Ireland indicated that it considered the former acceptable.

The issue eventually became primarily a dispute between France and the UK. France argued for the implementation of the access provisions in Regulation 2141/70, which meant 'fishing up to the beaches'. The UK argued for as wide exclusive limits as it could possibly obtain. Despite many hours of debate within the Council, the question was not resolved until the conservation policy was settled on 25 January 1983. It was almost six wasted years because the provisions finally adopted were very little different to those agreed in the Act of Accession of Denmark, Norway, Ireland and the UK, except for the inclusion of an area off the north east coast of Scotland in which fishing is subject to a limited number of licensed vessels, the so-called 'Shetland box' (Fig. 3.1).

Most of the negotiations which led to its conclusion took place not in the Council but in bilateral discussions between the two main protagonists, under the auspices of the Commission.

The allocation of resources: total allowable catches and quotas

Introduction

The second major issue which was to dominate the debate leading to the adoption of the conservation policy was the allocation of resources which realised itself as a debate about the division of total allowable catches (TACs) into national quotas. The origins of this debate lay in the question of access; if exclusive fishery zones could not be achieved then a second-best arrangement would be to have guaranteed fishing possibilities. The only universally accepted measure of fishing possibilities which could be divided in a

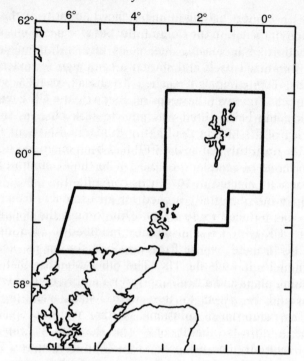

Fig 3.1 The Shetland box; the sensitive region within the meaning of Article 7 of Regulations No. 170/83 and No. 3760/92.

comprehensible and accepted manner was tonnages. The scientific community had long developed the models by which catch possibilities could be calculated for different rates of fishing mortality and by the 1970s had the long series of data required to use these models. Catch possibilities could be calculated for the most important stocks. The route was open for adopting a method of allocation which had the necessary scientific basis of conservation of the stocks while meeting the political requirements of allocating the resources. Additionally, TACs allocated by quotas had been recommended by both the International Commission for the North-west Atlantic Fisheries (ICNAF) and NEAFC so they were measures with which fisheries administrators were already familiar. However, there is a world of difference between agreeing a principle and agreeing a method by which to implement it. The debate about the method raged for over five years.

Relative stability

All concerned in the debate were determined that whatever method was agreed, it should be used for a long period. Neither the Commission nor the Member States wanted a system which would entail a long debate each year as to how the TACs were to be allocated. Additionally, all Member States wanted a system which guaranteed their industries specific tonnages which would allow them to plan their futures. This was despite the fact that the fishing industry had had to live from its birth with fluctuating catches which, until the advent of the means of calculating catch possibilities, had been totally unpredictable. The Commission was fully aware that it was impossible to guarantee specific tonnages because of the marked natural fluctuations in catch possibilities from one year to the next, resulting from natural causes outside man's control. For this reason it proposed that each Member State should be guaranteed only a specified percentage of the TAC for each stock, to which the term 'relative stability' was given. Relative stability was to be applied on a stock by stock basis. If it had applied to all TACs combined, it would have precipitated long and almost unresolvable arguments as to how a Member State was to be compensated for decreases in catch possibilities in those stocks for which it was allocated quotas. But this still left the question of determining a method for calculating relative stability.

Two of the factors which were to be taken into account had already been proposed by the Commission. One was catches based upon a reference period, to be known as 'historic catches'. The other was special provisions for areas heavily dependent upon fishing. The latter became known as the 'vital needs' of these Member States and later as the 'Hague Preferences' following the elaboration of what these preferences were to be in a declaration of the Council at a meeting in The Hague in October 1976. To these was to be added compensation for jurisdictional losses. The question of allocation of TACs broke new ground, each Member State attempting to find a method which would result in its receiving the greatest possible share.

One of the issues on which all Member States were agreed was that all the most important resources should be allocated. For the majority of fish stocks there was no scientific advice, based on analytical assessments, on which to propose TACs so the best available information was used; often the only information available was catches. These TACs came to be termed 'precautionary TACs', in contrast to those based upon analytical assessments.

Historic catches

The debate on historic catches was typical of so many within the Council. Those Member States which would have benefited from a long reference period argued for such a period; those who would have benefited from a shorter period argued the reverse. The Commission resolved this difficulty by proposing a period, 1973–78, which suited no-one and, as so often happens in such cases, was accepted for this reason. The year 1978 was selected as the end of the period because it was, at the time, the last year for which official catch statistics were available; the Commission was determined to use figures which were published, in the Bulletins Statistiques of the International Council for the Exploration of the Sea (ICES) and which could not therefore be disputed. Nor could a Member State conveniently 'update' its catch statistics. Even though the negotiations dragged on for years, the reference period was never revised because that would have put the negotiations back to their starting point.

Despite this determination to use official catch statistics, they were modified in two ways. Most importantly, all Member States, with the exception of

Denmark, were determined that Denmark should not benefit from its large by-catches of human consumption species caught in its fisheries for industrial species. The estimated quantities of these were therefore deducted. Second, the Commission had proposed TACs for 1978 which it had called upon Member States to respect. Catches in excess of these TACs were excluded, a means whereby the Commission could impress on the Council its dedication to conservation principles.

Hague Preferences

Neither the provisions for the Hague Preferences of certain coastal regions nor the areas to which they were to apply had been defined by the Council resolution. Again the Commission proposed what these should be both in quantitative terms and the areas, particularly in the definition of 'Northern Britain'.

For Ireland the quantities were twice its total landings for 1975 in all ports and, for the UK, landings by vessels of 24 metres or less in 1975 in Northern Ireland, Scotland and in ports on the north-east coast of England as far south as Bridlington inclusive. The quantities for Ireland ostensibly took account of an existing plan for the doubling of the Irish fleet. They also recognized the underdeveloped state of Irish fisheries at that time as well as the large Irish fisheries zone to which Ireland, like the UK, had wanted much more restricted access than it eventually was able to obtain. This was part of the price paid to Ireland in order to get its acceptance of the finally agreed solution to the question of 'access'. For the UK, it had initially been the intention to define 'Northern Britain' only as Northern Ireland and Scotland. The UK argued for the inclusion of the ports on the north-east coast of England as far south as, and including, Bridlington, an argument which the Commission accepted. Although other Member States strongly disputed the definitions, they finally agreed.

At that time Greenland was still part of the Community and also received Hague Preferences which approximated to the catching power of its domestic fleet.

Compensation for jurisdictional losses

The final element taken into account in the quota calculations was compensation for losses experienced by Community fleets, mainly those of Germany and the UK, caused by the extension of fishery limits to 200 miles by third countries. This factor entered into the negotiations only at a late stage. It was calculated as the difference between the average catch of each species for 1973–76 less the theoretical catch possibilities for each Member State in the waters of each third country. The theoretical catch possibilities were calculated by multiplying the TAC fixed by the third country by the percentage of the total catch of each species taken by the Member State in the period 1973–76 in those waters. Use of the TAC fixed by the third country took into account fluctuations in catch possibilities which would have affected the catches by each Member State even if there had been no change in fishery limits. It was, and still is, an axiom of the Commission that conservation losses, defined as reduction in catches caused either by the implementation of conservation measures or those arising from natural fluctuations in fish stocks, should never be compensated. The reason is obvious; compensation could only be given from other stocks fished by Member States other than those affected, leading to endless wrangling.

The mathematics of allocation

Having agreement upon the factors which were to be taken into account and applying these factors in a manner which produced acceptable results were two different things. Much ingenuity went into producing a method of calculation which had to be totally transparent if it was to be acceptable to the Member States. The application of the Hague Preferences proved a particularly difficult problem to resolve because it disadvantaged some Member States much more than others. A method had to be found which spread the 'costs' of the transfers to the beneficiaries evenly between the losing Member States. A similar but lesser problem arose over the compensation for losses in third country waters for which, in the final calculations, it was impossible to provide full compensation. All this took much ingenuity on the part of the Commission. The method is still largely not understood, which has repercussions on the perception of the operation of the CFP to this day. It is therefore worth describing in detail, complex though it is.

First, the species which were to be subject to TACs were divided into four groups. One group consisted of what were termed 'the seven main species': cod, haddock, saithe (coalfish), whiting, plaice, redfish and mackerel. A second group consisted of all the other species used for human consumption, except herring. The third group consisted of herring only and the fourth group of species used at that time almost entirely only for reduction to meal and oil: blue whiting, horse mackerel, Norway pout, sandeel and sprat. Herring was treated as a separate group because most of the stocks were collapsed or recovering from collapse in the 1970s. In particular there was no fishery for the potentially largest Community resource of herring, that in the North Sea, and therefore no basis on which to agree allocations. For this reason, the method of allocation was not applied to herring. Nor was it applied to the industrial species for the reason that most of these were caught by Denmark only. For the other two groups the method of calculation was as follows.

Within each group of species, the TAC for each stock of fish, recommended by ACFM, was multiplied by the percentage share of the Community catch by each Member State for the period 1973–78; in the absence of a recommended TAC the Community catch for 1981 was used. This calculation gave an allocation for each Member State. The calculation was done twice, first as already described and, second, by first allocating to those Member States which benefited from Hague Preferences the agreed quantities and then allocating the remainder between the other Member States. An example is given in Table 3.1.

In order to take account of the relative value of the different species, all

Table 3.1 Example of the method by which account was taken of Hague Preferences in allocating catch possibilities. (See text for further explanation.)

Stock: Cod VI

Member State	D	F	NL	B	UK	DK	IRL	EEC
(a) 1973–78 average catches	6	4304	6	62	8556	1	1025	13 960
(b) Line (a) as %	0.04	30.83	0.04	0.45	61.29	0.01	7.34	100.0
(c) 1973–78 average catches adjusted for Hague Preferences	5	3646	5	53	3528	1	0	6986
(d) Line (c) as %	0.07	52.33	0.07	0.76	46.76	0.01	0	100.0
(e) 1981 national allocations based on line (d)	2	1574	2	23	1407	0	0	3008
(f) 1981 national allocations, including Hague Preferences	2	1574	2	23	6117	0	2282	10 000
(g) Line (f) as %	0.02	15.74	0.02	0.23	61.17	0	22.82	100.0
(h) National allocations without Hague Preferences based on line (b)	4	3083	4	45	6129	1	734	10 000
(j) Hague transfer (line (f)—line (h))	− 2	− 1509	− 2	− 22	− 12	− 1	+ 1548	

Hague Preferences

	UK	IRL	
(1975 catches)	4710	1141	
	1975 catch	1975 catch × 2	Total
	4710	2282	13 960

D= Germany; F=France; NL= the Netherlands; B= Belgium; UK= United Kingdom; DK= Denmark; IRL= Ireland

Source: Commission document SEC (81) 105

allocations were converted into 'tonnes, cod equivalent'. This was a system which had been invented during the negotiations with the Faroe Islands, Norway and Sweden in order to permit exchanges of fish having approximately equal value. A series of factors by which actual tonnages were multiplied in order to convert the quantities which were being exchanged into equivalent tonnages was agreed. This was done by taking the value of cod as 1.0 and relating all other species to cod by a factor which represented approximately its relative market value. These factors were called 'cod equivalents' and tonnages, calculated using these factors, 'tonnes cod equivalent'. For example, haddock and plaice also have cod equivalents of 1.0 but that for saithe (coalfish) is 0.77 and for whiting 0.86 so an allocation of 100 t of saithe counted only as 0.77 t cod equivalent. In the reverse, in negotiations, 'selling' a tonne of cod 'buys' 1.3 t of saithe or 1.16 t of whiting.

Returning to the calculations, for each group of species the total tonnages were summed in two ways. One summation was of the total allocations for all stocks calculated without taking account of the Hague Preferences. The other summation was of the total allocations taking account of the allocation of Hague Preferences, but only if this gave the beneficiary Member States higher allocations. This was not always the case.

The differences between the two sets of figures for those Member States which lost through the application of Hague Preferences had to be 'equalized'. This was done by taking the total sum of the differences and allocating it between the Member States in proportion to their nominal catches. These nominal catches were calculated by taking the the total Community catch for 1981 and allocating it by the percentage catches for 1973–78. To make the calculation even more complicated, the UK had to be treated as two Member States because only 'Northern Britain' benefited from Hague Preferences. The same applied to Denmark in respect of Greenland. It was also applied to St Pierre and Miquelon even though France did not benefit from Hague Preferences for the reason that it was impractical to take fish from these two islands (Table 3.2).

The method of calculating jurisdictional losses has already been described. These were also 'equalized' by a procedure similar to that used for the Hague Preferences but, unlike them, jurisdictional losses were only partially allowed. The final result was a total allocation to each Member State of all the species in each group in tonnes cod equivalent.

It took many months and many Council meetings to agree on the method. Even more months were to pass before agreement was finally reached on 25 January 1983, the ministers rejecting one set of figures after another, each time setting a new deadline by which they promised themselves they would reach agreement. Meantime, the Commission juggled with the figures, updating them as new information became available. The crux of the allocation problem was that each Member State had already decided upon the percentage of the group of seven main species which it considered its 'fair share' of the resources and was determined to stick in its heels until it got it. Not surprisingly the total of their demands came to more than 100%. These percentages were inflated for negotiating purposes but it presented the ministers

Table 3.2 Example of equalization of Hague transfers. (See text for further explanation).
Species: cod, haddock, saithe, whiting, plaice, redfish and mackerel; in tonnes cod equivalent.

Line		EC	D	F		NL	B	UK		DK		IRL	
				St P & M	Mainland			North	South	Mainland	Greenland	Internal	External
1	Catch possibilities 1981 internal based on average catches 1973–78 + external catch possibilities	1 131 699	160 485	4 452	146 708	98 012	26 395	204 578	131 976	297 514	39 458	21 967	155
2	Hague transfers	± 45 233	– 13 703	+ 48	– 14 397	– 2 267	– 1 027	+ 10 761	– 10 004	– 3 835	+ 13 042	+ 21 318	0
3	Catch possibilities 1981 internal with Hague corrections + catch possibilities external 1981	1 131 699	146 782	4 500	132 311	95 745	25 368	215 339	121 972	293 679	52 500	43 348	155
4	Equalized Hague transfers (– 45 233/861 245) x (line 1)		– 8 429		– 7 705	– 5 148	– 1 386		– 6 931	– 15 626			– 8
5	Difference between lines 2 and 4		+ 5 274		+ 6 692	– 2 881	– 359		+ 3 073	– 11 791			– 8
6	Resultant total catch possibilities (lines 5 + 3)	1 131 699	152 056	143 503		92 864	25 009	340 384		334 384		43 495	

St P & M = St Pierre and Miquelon; other symbols as Table 3.1.
Source: Commission document SEC(81) 105.

with the problem that acceptance of any lower share represented a 'climb-down' in the eyes of its fishing industry. It also created a problem for the Commission. Every time it redid the calculations, using the latest data, the figures changed and the 'losers' found the new results unacceptable, a word which quickly becomes recognizable in all Community languages. The UK based its opening claim on its contribution to the resources, of the order of 60%. This was totally unrealistic in terms of its historic share of the catches, but it made it difficult to negotiate downwards to what all other Member States considered a reasonable percentage. Other Member States found such a high claim unacceptable for two reasons; first, it would have required major reductions in their own fisheries but second, and more importantly, it was totally contrary to the Community principle that everything which goes into the Community pool is allocated between Member States on the basis of their needs, not on the basis of 'getting out what you put in', as the UK also found several years later in its negotiations over budgetary contributions. How little the percentage allocations changed as a result of three years of intensive negotiation is shown by Table 3.3.

Table 3.3 Comparison of the proposal of the Commission for the allocation of the seven main species (cod, haddock, whiting, saithe, plaice, redfish and mackerel), in tonnes cod equivalent, in July 1980 with that adopted in January 1983.

Member State	July 1980		January 1983	
	tonnes cod equivalent	%	tonnes cod equivalent	%
Belgium	20 376	1.99	28 900	2.09
Denmark	269 285	26.25	317 900	22.99
France	123 912	12.08	182 800	13.22
Germany	130 597	12.73	184 900	13.37
Ireland	45 249	4.41	61 000	4.41
Netherlands	90 513	8.82	100 700	7.28
UK	345 926	33.72	506 700	36.64
Total	1 025 858		1 382 900	

Sources: Commission document COM(80) 452 and Council Regulation (EEC) No. 172/83.

Three factors were to facilitate agreement. The first of these has had a major impact on the conservation of the fish stocks. In the late 1970s and early 1980s the policy of ACFM was to recommend TACs which corresponded to reductions in fishing mortality rates of 10–20% from existing levels. The objective was gradually to reduce fishing mortality rates to the levels which would give the maximum yield from each young fish entering the fishery, F_{max} (see Chapter 8). It became obvious to the Commission that not only were the Member States not prepared to accept TACs which corresponded with such reductions but it made little practical sense to try to fix TACs at these levels because the fleets which were creating the fishing mortality rates were in existence. It would be far better to fix TACs which corresponded with existing fishing mortality rates and once having agreed a conservation policy, take steps to reduce both the size of the fishing fleets and the fishing mortality rates simultaneously. Doing this increased the catch possibilities considerably.

The second factor was that the Commission decided to propose TACs which permitted Hague Preferences to be met, even if this meant proposing TACs which were much higher than those recommended by ACFM. The basis for this decision was to avoid creating yet another reason for Member States not to agree on the conservation policy. This resulted in the TACs for the stocks of Irish Sea cod and whiting being fixed much higher than any possible levels of catches for the reason that, not only both Ireland and the UK have Hague Preferences in these two stocks, but it was essential to allocate quotas to other Member States which had an historic record of fishing in the Irish Sea. As already described in Chapter 2, this decision still has consequences for the policy in 1993.

The third factor was that, starting in 1964, there was a massive increase in catches of cod, haddock and whiting in the North Sea and, to a lesser extent, west of Scotland, to which the name 'gadoid outburst' was given. These increases reflected an increase in the number of young fish entering the fishery, for which no reason was evident. This was particularly marked for North Sea cod, one of the most sought after and therefore one of the key stocks in the negotiations. In 1974, the stock size was 551 000 t from which it increased year after year almost without a break, reaching its second highest recorded level of 881 000 t in 1980 (Fig. 3.2). The proposed TACs rose correspondingly.

The higher the actual tonnages available, the more the Member States were prepared to compromise on their target percentages. The increase in the quantities of fish available eventually led to the position where the percentage allocations on the table became acceptable but the deciding factor was that if agreement had not been reached by 31 December 1982 there would have been, in theory, fishing up to the beaches as from 1 January 1983 under Article 2 of

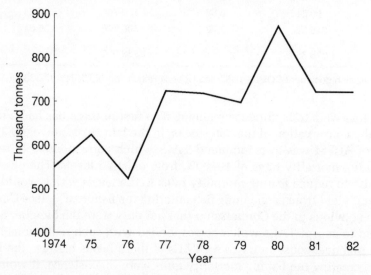

Fig. 3.2 The total stock size, in thousands of tonnes, of North Sea cod, 1975–82.

Regulation No. 101/76, which by now had replaced Regulation No. 2141/70. Kent Kirk, a well-known and flamboyant Danish fisherman and politician, who was later to become Minister for Fisheries, did try it when agreement was not reached in December 1982 when there was a last minute hitch caused by Denmark's refusal to accept the final proposal of the Commission. This was eventually resolved by allocating to Denmark additional quantities of North Sea cod for three years and making special provisions for mackerel in order to compensate for what Denmark considered its unjustified exclusion from fishing for western mackerel.

There is one final twist to this story which many appear not to realize ever occurred. Although agreement was based essentially upon the percentage allocations in the two groups, before final agreement could be reached the overall percentage allocations in tonnes, cod equivalent, for the two groups of species had to be converted back into actual tonnages of fish on a stock by stock basis. This final exercise gave each Member State the opportunity to increase its catch possibilities of those stocks of most importance to it but to do so it had to make sacrifices in others. For example, the UK was particularly interested in obtaining more than its historic catch of sole in the English and Bristol Channels and in the Celtic Sea, which it achieved at the expense of cod and plaice. Now, of course, English fishermen complain about their relatively small share of the cod TAC in the English Channel.

This procedure was not followed for herring. For those stocks for which fishing was permitted at the time of these settlements, allocations, based essentially upon historic catches, were negotiated.

The 'Luxembourg compromise'

Agreement might have been reached much earlier if the Council had not been operating under the provisions of the 'Luxembourg compromise' throughout the period of the negotiations. Community rules provide for a system of qualified majority voting, as described in Chapter 1. However, in order to resolve a problem created by France which had boycotted meetings during the latter half of 1965, the Council agreed at a meeting held in Luxembourg in 1966 that it would not vote on a proposal of the Commission if a Member State declared that to do so would affect its 'vital national interests'; these did not have to be specified. The device was continually used throughout the negotiation of the conservation policy to block agreement, often with a linkage being made to other negotiations, notably those with third countries, in order to prevent decisions from being made.

Technical conservation measures

'Technical conservation measures' is the term applied to measures which regulate the type of fishing gear used, such as minimum mesh sizes, the size of fish which may be landed, such as minimum mesh size and by-catch limits and seasonal restrictions on fishing, such as seasonal bans on fishing in nursery areas. Their general objective is to minimize catches of small fish.

The debate concerning technical conservation measures was never as heated as that concerning the allocation of resources, for two reasons. First, although technical conservation measures affect the livelihoods of fishermen, they involved no commitment to a principle such as relative stability. Second, all Member States were familiar with them, all having national legislation. Much of the debate concerned not the measures themselves but the competence of the Member States and the Community to make regulations. The resolution of this debate had fundamental consequences in determining that it is the Community, not the Member States, which has total competence for fisheries.

That the adoption of regulations concerning technical conservation measures was initially not controversial is demonstrated by the fact that the Council adopted 12 regulations in 1977. These concerned mainly bans on fishing for herring in the North Sea, Celtic Sea and west of Scotland and measures concerning fishing for Norway pout. The validity of none of these extended beyond 31 January 1978 except a regulation banning fishing for herring for purposes other than human consumption, which is still in force. Also, Member States were able to introduce unilateral conservation measures until 1 January 1979 when this power ceased. This change in competence stemmed from Article 102 of the Act of Accession, which read:

> 'From the sixth year after accession at the latest, the Council, acting on a proposal from the Commission, shall determine the conditions for fishing with a view to ensuring protection of the fishing grounds and the conservation of the biological resources of the sea.'

It was not immediately conceded, however, by the Member States, in particular the UK, which contested this interpretation by the Commission of this article. But it was confirmed by the Court of Justice in a decision in a case brought by France against the UK concerning fishing for *Nephrops norvegicus* (Norway lobsters) with small meshed nets. As a consequence of this and other rulings of the Court of Justice, competence in all fisheries matters was to pass to the Community. This example illustrates how the Commission acts in interpreting and implementing the treaties. However, with the Council unwilling to adopt conservation measures proposed by the Commission, the Community was paralysed. The Council therefore agreed by a series of declarations and resolutions that, prior to the adoption of a conservation policy, Member States could introduce conservation measures but only after requesting and obtaining the permission of the Commission. This history is described in detail in the books listed in the references.

The rulings of the Court of Justice led to the resolution of the one major problem, a bitter dispute between Denmark and the UK concerning fishing in the Norway pout box as well as the size of that box (Fig. 3.3). The area is not the only one in which Norway pout are abundant but is also the main nursery area in the North Sea for haddock and whiting. These are caught as a by-catch in the fishery for Norway pout, for which it is necessary to use nets having a small mesh. These by-catches are often very large. British fishermen claimed

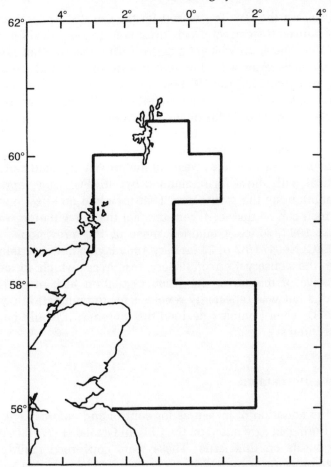

Fig. 3.3 The Norway pout box; the area of the North Sea in which it is prohibited under Article 2, Annex I, of Regulation No. 3094/86 to fish with nets less than standard for the area, thus preventing fishing for all industrial species but, in particular, Norway pout.

that they had a very detrimental effect on the catches of haddock and whiting taken in the fisheries for human consumption, a claim supported by the scientific analyses made by ICES. British fishermen argued, and still argue, that preference should be given to fishing for species used for human consumption and that the fishery for Norway pout should be reduced to the lowest possible level, preferably banned altogether. Denmark argued that there is no alternative use for Norway pout and not to fish it would be a waste of a resource.

In 1978 the UK informed the Commission that it intended to extend the area of the Norway pout box in which fishing for Norway pout would be banned. This legislation did not conform to proposals which the Commission had already made. Also, the Commission did not consider that the UK had followed the correct procedure in asking approval for its measure. The Commission therefore introduced a case against the UK in the Court of Justice which ruled against the UK. The reason for the Court's decision was that the UK had not followed the correct procedures in asking approval for its measure but also that the legislation was not a conservation measure but one designed to protect British fisheries and therefore discriminatory and contrary to the Treaty. This ruling decided the UK that it was essential to resolve this problem with Denmark, which was done mainly by bilateral contacts under the auspices of the Commission. This opened the way for the adoption of the package of technical conservation measures.

Apart from this issue, the main reason for which the technical conservation measures provoked few problems was that the provisions finally adopted were almost identical with those recommended by NEAFC. They represented no more than maintaining the *status quo*. That they had no effect on the activities of fishermen can be deduced from the fact that a regulation was adopted in September 1980 which contained most of the provisions of Council Regulation (EEC) No. 171/83 of 25 January 1983 laying down certain technical measures for the conservation of fishery resources, which formed part of the CFP package of measures. The 1980 regulation had an initial validity of three months but was repeatedly renewed for three-monthly periods until 31 October 1981 when Denmark decided that to renew it would be against its vital national interests.

Control and enforcement

For the UK, the adoption of measures for control and enforcement was a key issue in concluding an agreement on the CFP. In October 1977 the Commission made its proposals on this issue. These were concerned mainly with the administrative procedures by which catches were to be recorded and reported to the Commission in order that the uptake of quotas could be monitored and which the Commission could follow if it considered that Member States were not carrying out enforcement effectively.

The proposals caused little debate, except on three issues. First, the Commission wanted all species caught recorded in the logbooks. Member States argued that the logbook was a means of enforcing quotas, not of collecting scientific data. Fishermen should have to do no more than was necessary to collect the data required for control purposes; this argument won the day. Second, the Commission proposed that landings should be made only in recognized ports, in order to facilitate control. The Member States argued that this was an unwarranted restriction on the freedom of fishermen and again won the argument. Third, a provision for fishing plans was also opposed by the Member States and did not feature in the regulation as adopted.

The key issue, that of responsibility for control and enforcement, had already been conceded in the proposal of the Commission. This was left the responsibility of Member States. The Commission was provided with certain powers but these were limited. The regulation provided for Commission inspectors but they had no legal powers; all they could do was to observe and report. Their programmes of inspection had to be 'mutually acceptable' and agreed beforehand and were hedged around with many restrictions. Even the provisions by which the Commission could intervene if it considered that a Member State was not carrying out its responsibilities adequately left the Commission in a weak position. The relevant provisions read:

> 'If the Commission considers that irregularities have occurred in the implementation of this Regulation, it shall inform the Member State or States concerned, which shall then conduct an administrative inquiry in which Commission officials may participate. The Member State or States concerned shall inform the Commission of the progress and results of the inquiry.'

It is significant that, as with the initial regulation on technical conservation measures, the Council adopted the first control regulation (Regulation No. 2057/82) on 29 June 1982 in advance of the CFP package. If it had included provisions which would have made for effective enforcement, this would never have happened.

The agreement of 25 January 1983

All the threads of the CFP were finally drawn together on 25 January 1983 with the adoption of 12 regulations. The most important of these was Regulation No. 170/83 establishing a Community system for the conservation and management of fishery resources, the basic regulation for the conservation policy. It provided the basis for TACs and quotas, relative stability and the continuation of the provisions on access, although in a modified form to those included in the Act of Accession of Denmark, Norway, Ireland and the UK. The main difference on access was the provision for an area off the east of the Shetland Islands where fishing was subject to a system of limited licensing, 'the Shetland box'. The regulation also provided for a review of these provisions after 10 and 20 years.

Regulation No. 171/83 laid down technical measures for the conservation of fishery resources while Regulation Nos 172–181/83 laid down the TACs and quotas for 1982 for Community vessels fishing in the waters of the Community, the North-west Atlantic Fisheries Organization (NAFO), Canada, Jan Mayen, Sweden, the Faroe Islands and Norway as well as the reciprocal provisions. On the face of it, to adopt TACs and quotas for a year that had already ended was ludicrous but these were a vital part of the agreement; the percentage quotas laid down in these regulations fixed relative stability.

There was much euphoria and self-congratulation amongst those present at this historic Council meeting as the champagne glasses clinked and all congratulated each other. All believed that the basis had been set for a prosperous Community fishing industry based upon rational management of the stocks in balance with fleet capacity, managed by a structures policy, and regulated markets. How differently events were to evolve!

Chapter 4

Adopting Total Allowable Catches and Quotas

or The House of Cards

The House of Cards

The history of the operation of the CFP falls not only into separate phases but also into separate fields of political interest. From 1970 until 1983 only the markets and structures policies existed. From 1977 onwards the policy for international fisheries relations came into being but it was not until the adoption of the conservation policy in 1983 that the CFP was completed.

Until approximately 1986 the four policies were implemented almost independently. To some extent this was an accident of history. Prior to the establishment of the Community it was in the national interest of countries to subsidize their fishing fleets through national markets and structural policies irrespective of the effects which this might have on the conservation of the fish stocks. There was no point in the era prior to the introduction of 200-mile fishery limits in a few nations taking conservation measures if other nations fishing the same stocks did not do so. This attitude of operating policies independently not unnaturally carried over into the Commission when the Community was founded. Besides, fishery limits were not extended to 200 miles until 1977 and there was no Community conservation policy until 1983. In these circumstances there was no pressure to consider the effects on a policy which did not yet exist. Furthermore, the Community had a deficit in supplies in all species caught in its waters, except for anchovy and sardines and, as already described, the main objective of the structural policy was to remedy this deficit by granting subsidies for vessel construction to catch more fish. Almost since their inception, the structural and market policies have caused few problems, at least in comparison with those experienced with the conservation policy, essentially for the reason that they provide financial support to the fishing industries of all Member States, which facilitates agreements.

Similarly, the policy for international fisheries relations has not caused major political debate. True, there have been periods when the non-renewal of an agreement before the expiry of an existing agreement, as has twice occurred with Morocco, has caused serious problems for those fishermen affected. But settlement has been reached reasonably quickly and, in the case of the first breakdown with Morocco, fishermen were compensated for loss of earnings. The question of fisheries was also a major point of debate in the negotiations leading to the withdrawal of Greenland from the Community as it was in the negotiations with the EFTA countries in the adoption of the agreement forming a European Economic Area but again both these problems were resolved relatively quickly. This policy also is not further discussed in this chapter. It has been the conservation policy which has roasted on the front-burner and caused most of the controversy. This and the next chapter tell that story.

The settlement of the CFP on the night of 25 January 1983 can be likened to the end of a war of attrition. All the parties had been exhausted by the long years of negotiations and all were thankful that it was now over. However, there was an acute awareness that the CFP represented an uneasy compromise; within the Commission, the settlement was likened to a 'house of cards' which could easily tumble if any party tried to move one of the cards. It was considered vital that no Member State should be provided with any opportunity to do so. The initial objective of the Commission was, therefore, to establish a routine under which all Member States would consider that it was 'normal' to adopt conservation measures, in particular TACs and quotas, without raising issues of principle. Admirable as was this objective initially, it led to almost total rigidity in the implementation of the conservation policy with the result that the Community failed to tackle the problems of overfishing which the settlement of 25 January 1983 had not attempted to resolve; as described in Chapter 3, the TACs were fixed on the basis of existing rates of fishing mortality, not on their reduction.

Additionally, a series of major political issues arose which totally absorbed the attention of the Council and the Commission, further preventing them from addressing the conservation issues. Events during the years 1990, 1991 and 1992 have shown that the 'house of cards' is far more stable than the Commission originally thought. With hindsight, progress might have been possible earlier, thus avoiding some of the traumas experienced in 1991 and 1992 although, given the Community's unwillingness to take action unless forced upon it, this is perhaps an over-optimistic view.

There are three main threads to the story: (1) TACs and quotas, (2) technical conservation measures and conditions of fishing (the technical conservation measures which form part of the annual TAC and quota regulation) which are for this reason treated together, and (3) control and enforcement. This chapter tells the history of the implementation of TACs and quotas.

The history of TACs and quotas can be divided into three periods, 1983–1986, 1987–89 and 1990 onwards. The first of these periods was one in which the prime objective of the Commission was to propose nothing which might result in the newly agreed policy unravelling. The second period was characterized by seemingly endless debate over one issue after another. The issues were politically very important but they prevented any attempt being made to advance the conservation objectives of the policy. The third period is, because it still continues, a period in which an attempt has been made to address these issues.

The period 1983–86

Establishing a routine

The TACs and quotas which were adopted on 25 January 1983 were those for 1982 so the immediate objective of the Commission was to propose and get adopted the TACs and quotas for 1983 as quickly as possible. This was seen as a first step in getting TACs and quotas adopted in the year prior to that to which they applied, thus establishing a routine which all Member States would come to regard as the 'normal' pattern. Precedent plays a very important part in the proceedings of the Community; once something has been done a first time, even if at the time it is stated that it is an *ad hoc* solution, it very often then becomes the established method of operation. In this case, once TACs and quotas had been adopted for the first time prior to the end of the year to which they applied, the Council would feel obliged to maintain the pattern, both because their fishing industries would expect it of them and because the ministers would regard it as a political failure not to do so. This has proved the case.

There were other reasons for getting the TACs and quotas for 1983 adopted quickly: their adoption would be the first test of the acceptance of the principle of 'relative stability'; getting them adopted would confirm the principle; and getting them adopted quickly would reduce the time in which Member States might have second thoughts about the agreed allocations.

In the end it took almost a whole year to agree the TACs and quotas for 1983, the regulation not being adopted until 20 December 1983. However, its adoption confirmed the principle of relative stability – there was only one change from the allocations agreed for 1982. Belgium obtained an allocation for the management unit 'west of Scotland and Rockall haddock' by arguing that it was illogical for it to have quotas for other stocks in this area which it could not fish without catching haddock; without a quota it would be forced to discard them, which would not result in conservation of the stocks. For once, logic, helped by the fact that Belgium is a small country, won the day. This is the only permanent change in the original allocation keys which has ever been made.

Allocation of North Sea herring

There was a reason for the delay in adopting TACs and quotas for 1983. In adopting those for 1982, there remained one very important card to add to the 'house of cards', the allocation of North Sea herring. Following the collapse of the stock of North Sea herring, there had been an effective Community ban on fishing herring in the North Sea since September 1977. This was implemented by a ban on fishing herring for reduction to meal and oil but as there were insufficient adult herring to make a fishery for human consumption purposes economically viable, this amounted effectively to a total ban on fishing for herring. This regulation still remains in force. With no catch possibilities, the Commission had tried to reach agreement on the basis of possible TACs once the stock had recovered, using proposals based upon what were termed 'artificial herring', but the Member States were not prepared to agree allocations on this basis. They wanted to know the catch possibilities in actual herring.

In 1983, the stock was starting to recover so the debate concerning TACs and quotas for 1983 took place in parallel with that about the allocation of North Sea herring. The debate was complex for two reasons: first, the North Sea stock of herring is jointly managed with Norway, with whom the Community had yet to establish an allocation key; second, the pattern of catches by both area and consequently by Member State had changed markedly over the years as the area where the largest catches had been taken had moved progressively from the southern to the middle and finally to the northern North Sea. The historic reference period of 1973–78 which had been acceptable for other stocks was not acceptable for herring.

After a series of very acrimonious meetings between the Commission and the Member States an allocation key was finally adopted by the Council at its meeting of 20 December 1983. In order to reach agreement an allocation key, which changed as the quantity of herring available to the Community varied, was adopted instead of the fixed keys which applied to all other stocks. The adoption of this allocation key permitted the adoption of the TACs and quotas for 1983 at the same time. This paved the way for the rapid adoption of TACs and quotas for 1984 on 31 January 1984 with, finally, the achievement of the Community's objective, the adoption of those for 1985 on 19 December 1984.

Having achieved this objective, the Community was not able to relax because winds now started to blow, first from one direction and then another, which threatened to destroy 'the house of cards'. Throughout 1984 the wind of North Sea herring threatened the fragile structure, with Member States seeking a far higher TAC than the Commission was prepared to propose in the light of the scientific advice. The wind did not eventually die down until the scientific advice permitted the Council to propose an autonomous TAC of 155 000 t which was adopted by the Council at its meeting on 7 June 1984, thus satisfying the demands of the Member States for the recommencement of fishing for North Sea herring on a major scale. However, this TAC was fixed unilaterally without any agreement with Norway on an allocation key. This key was not agreed until 1986 for exactly the same reasons that the Member States found it difficult to agree an allocation key amongst themselves. This is a wind which has continued to gust and eddy. Norway is dissatisfied with the

Table 4.1 Agreed allocation key for North Sea herring between Norway and the EEC.

Spawning stock biomass (million tonnes)	Percentage share of TAC	
	Norway	EEC
1.0 – 1.499	25	75
1.5 – 1.999	29	71
2.0 – 2.5	32	68

key which was agreed (Table 4.1). Also, Norway has been able to use complaints of overfishing of the Community share of the TAC, always strenuously denied by the Community, as a means of obtaining in the course of the annual negotiations, a higher share of the TAC than that to which it is entitled on the basis of a straight application of the key, much to the anger of the Member States.

The accession of Spain and Portugal

The next depression, in more than one sense of the word, which was developing was the accession of Spain and Portugal on 1 January 1986. It was essential for the Community to adopt TACs and quotas for 1986 for the existing ten Member States prior to their accession in order that they should form part of the *acquis communitaire* which Spain and Portugal would be obliged to accept. Otherwise the two new Member States would have been able to join in the debate with unpredictable consequences. On this occasion neither Ireland nor the UK demurred, in contrast to when they acceded and were subject to similar treatment. However, fixing TACs only affected one year and was not permanent, as are the provisions concerning access. The Commission therefore made a proposal which could be adopted with the minimum of difficulties. The objective was achieved, the TACs and quota regulation for 1986 being adopted on 20 December 1985. This regulation was subsequently modified to allow for the addition of new stocks consequent upon the accession of Spain and Portugal. This necessitated modifying the allocation keys for stocks already subject to TAC for which the two new Member States obtained quotas. This modification was adopted easily and painlessly on 31 December 1985 but only at the cost of fixing for most stocks TACs which far exceeded historic catches. Unfortunately, the articles in the Act of Accession of Spain and Portugal were so badly drafted that they are possible to interpret in many different ways. This gave rise to continuous difficulties which were not resolved finally, it is hoped, until the regulation which fixed TACs and quotas for 1991. However, the new allocation keys which were agreed for 1986 have remained largely unaltered since then, certainly for the most important stocks, confirming the Community's attachment to the principle of 'relative stability'.

The period 1987–89

Svalbard cod

No sooner had this depression passed than new storm clouds began to gather, this time concerning cod at Svalbard. This posed several problems, the first of which was the competence of the Community concerning fisheries in the waters in the Svalbard area. Management of the natural resources in this region is regulated by the terms of the Treaty of Paris, which was signed in 1920. All the signatories of this Treaty have the right to exploit the natural resources but Norway has the responsibility for their management. In fisheries, this means that the Treaty of Paris determines how the stocks should be exploited, but the regulations must be non-discriminatory which excludes TACs because quotas discriminate between fishermen of different countries.

Several Member States are signatories of the Treaty of Paris which led the Commission to claim that, with the assumption of competence by the Community for fisheries, the Community also subsumed the rights and responsibilities of the signatories of the Treaty of Paris. This was disputed by the relevant Member States but they had effectively conceded that battle when the Council adopted a zero TAC for capelin in the waters around Svalbard in the TAC and quota regulation for 1982 and continued to do so in subsequent years. That TAC was adopted in order to prevent Denmark developing an industrial fishery for capelin in the area, which might have had adverse effects on cod. All other Member States, being interested in cod and not at all in capelin, voted for the zero TAC and conceded the argument about competence incidentally. Such are the workings of the Community.

Returning to the story of cod, the cod which are found in the Svalbard area form part of the North-east Arctic stock of cod. This stock inhabits mainly the fisheries zones of Norway and the USSR, which jointly fix the TAC. The Svalbard area is one of the main nursery areas for the stock. Both Norway and the USSR have a major interest in the stock but it is vitally important to Norway because the cod spawn off the Lofoten Islands, northern Norway, where the fishery provides the main source of income for the population. Without the fishery, emigration from the area would be even higher than it is. In the early 1980s, even more so than now, it was strategically important to maintain the population in northern Norway because of the tension at that time between the western bloc countries and the former USSR. During the period 1980–86 recruitment to the North–east Arctic stock of cod fell to a low level and Norway became very concerned that high levels of fishing on the juveniles would impede its recovery. Much of this fishing was done by Spain, about which Norway had been able to do very little except prohibit Spanish vessels from fishing in its fishery zone. With the accession of Spain to the Community, Norway was able to apply pressure on the Community in the context of its overall fisheries relations. But the problem was complicated by the fact that Norway is responsible only for management, not control of the resources, which precluded direct negotiations between the Commission and Norway. Any negotiations about TACs in the area should have taken place at a meeting of the signatories of the Treaty of Paris but as these included Japan and South Africa, the Community did not wish to proceed down that path. Instead, the Commission proposed an autonomous TAC for the Community based upon the Community's historic percentage share of the catches from the stock. Fortuitously, the percentage coincided with the Norwegian figure and and the resultant autonomous TAC with the tonnage which Norway and the USSR had agreed should be caught in the Svalbard area.

After much bitter dispute, the Council adopted this proposal on 18

December 1986 as one of the TACs for 1987, but without fixing its allocation and with the provision that it could not be fished until the allocation had been agreed. Agreeing the allocation presented identical problems to those experienced with North Sea herring, each Member State having a different pattern of catches over the years. Also, some Member States had under-reported their catches and now wanted allowance made for this in the calculations. Each Member State argued, as usual, for the period which produced for it the most favourable result. The position was further complicated by the fact that those Member States which had refrained from fishing in the area when the stocks were low in order, as they argued, to conserve the stock (but more probably for the economic reason that they could get bigger catches elsewhere), piously argued that they should not now be penalized for having contributed to the recovery of the stock. On the contrary, they argued that they should benefit. This cut no ice with those who had continued to fish but neither did unreported catches cut any ice with the Commission which went back to the roots of the CFP, holding that it would take account only of officially reported catch statistics. In the end even this holy writ was slightly adapted in order to reach agreement. As for North Sea herring, an allocation key which varied according to the catch possibilities available was proposed. Faced with the fact that the Member States could not fish the TAC until it had been allocated, the Council eventually adopted the allocation key and the relevant quotas on 18 May 1987. However, 18 months had been consumed in wrestling with what was a relatively minor problem in the overall context of the conservation policy. In 1987, 21 000 t of cod represented less than 1% of the Community's catch possibilities subject to TAC and quota, calculated in terms of 'tonnes cod equivalent'.

North Sea cod and haddock and Hague Preferences

While the attention of the Community was concentrated on these issues, storms were brewing elsewhere. The cloud which had been hanging on the horizon since 1986 was the decline in abundance of the stock of North Sea cod, the maximum TAC recommended by ACFM having fallen from 259 000 t in 1985 to 130 000 t in 1986. This stock was, and still is, heavily exploited, only about 35% of the fish surviving from one year to the next. From 1986 onwards ACFM strongly recommended TACs which corresponded with large reductions in the rate of fishing mortality but the Council refused to accept these recommendations, fixing TACs considerably in excess of the maximum recommended TACs from 1986 onwards. But the storm did not eventually break over North Sea cod but over North Sea haddock when ACFM recommended a maximum TAC for 1988 of 120 000 t. With an agreed EC–Norway share of 77:23, this meant a Community share of the TAC of 92 400 t plus any haddock which the Community could 'buy' from Norway in exchange for other species. However, it was improbable that a total EC share could be obtained which equalled 99 883 t, which was necessary to meet the UK's Hague Preference. The basis for the maintenance of Hague Preferences is to be found in the preamble to the basic conservation regulation which states:

> 'Whereas, in other respects, that stability, given the temporary biological situation of the stocks, must safeguard the particular needs of regions where local populations are especially dependent on fisheries and related industries as decided by the Council in its resolution of 3 November 1976, and in particular Annex VII thereto.'

The Council meeting of 3 November 1976 was that held in The Hague at which the 'regions where local populations are especially dependent on fisheries and related industries' had been defined, thus leading to the determination of Hague Preferences (Chapter 3). The UK invoked the principle of maintaining Hague Preferences which it considers to be one of the fundamental elements of the CFP. Only Ireland, the only other Member State benefiting from Hague Preferences, supported the UK and even then, not vigorously. All the other Member States argued that account had been taken of Hague Preferences once and for all during the negotiations leading to the settlement of the conservation policy and that the UK could not now 'have its cake and eat it too'. The Commission sided with the interpretation given to the preamble by the UK. Eventually the problem was resolved by agreeing with Norway a TAC of 140 000 t which, together with 'purchases' from Norway gave an EC share of the TAC of 129 500 t, well above the minimum needed to meet the Hague Preference of the UK.

The recommended TAC of 185 000 t for 1988 presented no problems but in 1988 the storm burst with a vengeance when ACFM recommended a TAC of 68 000 t, accompanied by dire warnings about the high level of exploitation of the stock, the high rates of discarding and the decline in the size of the spawning stock biomass. Faced with this advice, there was no possibility of fixing a TAC which would permit the Hague Preference of the UK for North Sea haddock to be respected. A TAC of 68 000 t was agreed with Norway and proposed by the Commission. The UK did not disagree with the TAC or the resultant Community share of 62 500 t as long as its Hague Preference of 60 000 t was respected. This position completely ignored the fact that the catch possibilities calculated for North Sea haddock include the by-catches taken in the industrial fishery, the only stock except for North Sea whiting for which they do. Provision had to be made for these catches (2 000 t) which would have left 500 t to be shared between the other Member States. As incidental catches in their roundfish fisheries would have been much greater than 500 t, all the excess would have to be discarded with no conservation effect. Even as a negotiating position it was unreasonable. The UK even tried to argue that its share should be greater than 60 000 t, using part of the calculations made in deriving the quotas at the time of the settlement of the conservation policy, even though these were no longer relevant and formed only part of the complex of these calculations. (The UK did not, however, persist with this latter argument.) There were heated debates on the issue in the meetings of the Internal Working Group and COREPER as well as much behind the scenes discussion between the Commission and the Member States.

The Commission played its cards very carefully. In its initial proposal for TACs and quotas the Commission made no proposal for the allocation of North Sea haddock. With no figures to argue about, fishermen were largely deprived of the opportunity to apply pressure on their ministers. The customary series of bilateral meetings with Member States allowed the Commission to assess the latest situation. The Commission then made its proposal. The quotas for North Sea haddock were an average of the results obtained by making the calculations respecting and not respecting the UK Hague Preference. The proposal was presented as an *ad hoc* arrangement for 1989 only. The UK was cornered; for the UK not to accept the proposal on haddock, it would have had to vote against the whole proposal, which given the probable balance of voting might have resulted in the compromise failing to achieve a qualified majority. The UK did not want to be held responsible for this and accepted. As described in Chapter 1, the tactic of presenting a compromise package which must be accepted or rejected as a whole is very commonly used. On this occasion it was the UK which had to concede. On many other occasions it has been other Member States.

The *ad hoc* solution did not respect the principle of relative stability, much to the chagrin of the UK. But it is important to note that the *ad hoc* solution did not represent an abandonment of the principle. The solution was simply an unavoidable consequence of the state of the stocks. A pragmatic decision was needed to solve a potentially difficult problem. That decision was taken.

The issue also raised the question of by-catches which causes much resentment amongst British fishermen who appear not to understand why provision is made in the annual TAC and quota regulation for certain tonnages of North Sea haddock and whiting to be set aside to be taken as by-catches in the industrial fishery for Norway pout. They query why these tonnages are not added to the quotas for human consumption, giving preference to the fisheries for human consumption over those for industrial purposes. These are the only two stocks for which the calculations made by ACFM include the by-catches taken in the industrial fishery for Norway pout. Most of the fish taken in the industrial fishery are well below the legal minimum landing size and, in fact, should not be retained by the legal minimum mesh size which applies to the human consumption fishery. Whatever the rights or wrongs of the industrial fishery, the by-catches allowed would not be available to the human consumption fishery even if the industrial fishery were banned. The quotas for human consumption must, therefore, exclude them.

What had hopefully been adopted as an *ad hoc* solution for one year for one stock proved to be a precedent which started changing the CFP. For 1990, the maximum TAC recommended by ACFM for North Sea haddock was even lower than for 1989, only 50 000 t, making it necessary to use the *ad hoc* solution once again. Worse still, the maximum recommended TAC for North Sea cod tumbled below the level needed to meet the UK's Hague Preference of 43 179 t, resulting in the application of the same method to that stock also. The scientific advice also made it necessary to apply the *ad hoc* solution to west of Scotland whiting. This was one of several stocks for which, in order to avoid a potentially difficult issue at the time of the settlement of the CFP, the Commission proposed and the Council adopted TACs higher than those recommended by the scientific advice in order that the Hague Preferences of Ireland or the UK or both could be respected, as described in Chapter 3. For west of Scotland whiting a minimum TAC of 16 400 t is needed to meet the UK's Hague Preference; a TAC of 10 000 t was adopted, the *ad hoc* solution being used. In 1991 the *ad hoc* solution had to be applied once again to North Sea cod and haddock and west of Scotland whiting but this time the Commission decided to go even further. Grasping firmly the nettle of Hague Preferences, the Commission proposed TACs strictly in accordance with the scientific advice, irrespective of whether or not Hague Preferences would be respected. In the final compromise which was adopted, most of the initial proposals of the Commission were increased, some obviously to levels which enabled the Hague Preferences to be respected but for two stocks, Irish Sea cod and whiting, to less than the minimum levels. The 'house of cards' was thoroughly shaken but withstood the winds of change.

Western mackerel: 'flexibility'

During this period another issue arose to occupy those involved with fisheries in the Community, the question of what was to become termed 'flexibility'. This involved the stock of western mackerel, a stock which had given rise to problems even during the negotiations leading up to the settlement of the CFP. One of the main reasons for which Denmark voted against the settlement package in December 1982 was that it excluded Denmark from fishing this stock. Only by making special provisions for Denmark concerning mackerel, amongst other concessions, was the vote of Denmark obtained.

Another problem related to this stock was that it migrates into Norwegian waters, for which reason Norway claims joint management. This has always been rejected by the Community. In the early 1980s, the Community had good grounds for rejecting joint management because the proportion of the catches taken in Norwegian waters was so small, approximately 3–4%. But there was a major political reason for rejecting joint management; the size of the TAC for this stock plays a key part in permitting the adoption of the TAC and quota proposals each year. For this reason the Community wants to be able to fix it autonomously. This presented no problem until the migration pattern of the western mackerel stock started to change dramatically from 1981 onwards, as described by Lockwood in his Buckland lecture (*The Mackerel* published by Fishing News Books). Instead of spending almost its whole life history in Community waters west of the British Isles, it started to move increasingly northwards and eastwards, occurring in large quantities off the Norwegian coast in summer and migrating south through the northern North Sea in the autumn, whereas earlier it had occurred west of Scotland during that season.

For British fishermen the issue was simple; in fact, for them, there was no issue. The UK has a large quota for western mackerel; the mackerel off the north east of Scotland, in the North Sea, are undeniably western mackerel in the biological sense of the stock, a fact disputed by no-one. Therefore British fishermen were entitled to fish it; fishing it in the North Sea rather than west of Scotland made no difference to the conservation of the stock. This argument ignored one very important fact which is that, for the purposes of TACs and quotas, a stock is legally defined by area and species; its biological stock definition is irrelevant. If this were not so, enforcement of the legislation would be impracticable because it is impossible to tell to which biological stock an individual fish belongs. However, this cut no ice with British fishermen who started to fish mackerel in the North Sea and report it as caught west of Scotland in order that it should be counted against the UK quota for western mackerel.

The UK authorities tried to enforce the regulation but experienced great difficulty because the fishermen considered that they had right on their side. Having experienced very severe enforcement problems during the 1987–88 fishery, the UK approached the Commission for a solution which provided for a proportion of the quotas of western mackerel to be caught in the North Sea, a concept which became called 'flexibility'. However, this was easier to suggest than to achieve. First, most Member States, in particular Denmark, were opposed to the British suggestion. Second, and much more seriously, Norway also adhered to the doctrine that all mackerel found in the North Sea formed part of the North Sea stock of mackerel. The Community and Norway had already agreed that it was a joint stock and, therefore, any management measures had to be agreed by the two parties. The Community did not want to enter into negotiations with Norway because it was strongly pressing its claim to joint management and a large share of the stock as a result of the changed migration pattern of western mackerel. To have entered into negotiations would have strengthened Norway's position.

The Commission was sympathetic, largely because it recognized that the CFP could not be managed in a manner which appeared to fly in the face of reality. However, it was a very delicate issue and the Commission had to ensure that it had the support of sufficient Member States before it dared make a proposal. It was not in a position to do so when it presented its proposals for TACs and quotas for 1988, much to the fury of the British fishing industry, a fury intensified by the fact that the Commission continued to ensure that the legislation was enforced, as was its responsibility under the Treaty.

In the autumn of 1988, with the approach of the mackerel fishery, the debate was reopened. This time the Commission was able to convince enough Member States to support the idea of 'flexibility', even without much enthusiasm, and in the annual round of consultations with Norway obtained its agreement without having to concede joint management. As for its proposals on North Sea haddock, the Commission waited until the meeting of ministers to place its ideas on flexibility on the table, thus ensuring that ministers could not be influenced by their fishing industries. This time the tactic worked to the advantage of the UK and the proposal was adopted. The same provisions were proposed and adopted for 1990 and 1991 although for 1991 they did not appear in the original proposal of the Commission, either indicating a change of position on the part of the Commission or its use as a negotiating ploy to achieve a concession on the part of the UK on some other aspect of the TAC and quota regulation.

The period 1990–91

The start of this period was characterized by the Commission proposing TACs which were strictly in accordance with the scientific advice. Two people catalysed this event. First, Vice-President Marin was appointed commissioner for fisheries in 1989 and, second, Dr Serchuk was appointed the new chairman of ACFM in 1990. Following the appointment of Dr Serchuk, ACFM modified the manner in which it gave its advice for the stocks of North Sea and west of Scotland cod, haddock and whiting. Starting in 1991, instead of recommending TACs which corresponded with specific reductions in rates of fishing mortality, as it had done previously, it recommended that fishing effort on these stocks should be reduced by 30% in view of the serious decline in the spawning stock biomasses of the stocks of cod and haddock. Even though the spawning stock biomasses of the stocks of whiting were causing no concern, it was recommended that the same reduction in fishing effort should apply to these stocks because they were caught with cod and haddock.

Vice-President Marin was determined that the fisheries policy should take a new path, one in which the scientific advice was rigorously followed. For 1991 the Commission proposed TACs which corresponded exactly or nearly with the scientific advice; that for North Sea cod was even less. However, the Commission proved no more successful in 1991 than in previous years, the TACs adopted being higher than proposed, with one exception, that of North Sea sole (Table 4.2). But the Commission did achieve one breakthrough.

As described in Chapter 3, in order to facilitate agreement on the allocation of quotas and thus to get agreement on the conservation policy in 1982, the Commission had proposed TACs for many stocks in the region west of Scotland and the Irish Sea higher than the scientific recommendations in order to maintain the resultant Irish and UK quotas at or above their Hague Preferences. The Commission continued to evade this issue in subsequent years, proposing sufficiently large TACs which the Council adopted. The most notable examples were Irish Sea cod and whiting for which the TACs were fixed at or above 15 000 t for cod and 18 170 t for whiting. For other stocks this solution did not have to be adopted so consistently because the scientific advice often permitted the fixing of TACs which resulted in quotas greater than Hague Preferences. In 1991 the Commission decided to confront this issue. Even though the Commission was not totally successful in getting TACs fixed for 1991 which corresponded with the scientific advice, they were fixed at a much lower level than was required to meet the Hague Preferences (Table 4.2); a comparison of its proposed TACs for the stocks of west of Scotland haddock and Irish Sea plaice with those adopted indicates that they had to be increased to levels which permitted the Hague Preferences to be respected (Table 4.2).

In 1992 the Commission proposed TACs for the stocks of North Sea whiting and west of Scotland cod, haddock and whiting which were higher than those corresponding to the advised 30% reduction in fishing effort. Most of these were adopted unchanged, or slightly increased, by the Council (Table 4.3). But the TACs for all the stocks of plaice and sole adopted by the Council were

Table 4.2 Recommendations of ACFM, the TACs proposed by the Commission and those adopted by the Council for 1991 for selected stocks for which analytical assessments available and the minimum TACs needed to meet Hague Preferences, where applicable, in tonnes.

Stock	Recommended by ACFM	Proposed by Commission	Fixed by Council	Minimum TACs to meet Hague Preferences
Cod IV	100 000*	92 000	100 000	110 723 †
Cod VI	14 500*	14 500	16 000	18 430
Cod VIIa	6 000	6 000	10 000	15 000
Haddock IV	48 000*	48 500	50 000	99 883 †
Haddock VI	13 000*	13 000	15 200	15 180
Whiting IV	134 000*	135 000	141 000	61 274 †
Whiting VI	7 500*	7 500	9 000	16 400
Whiting VIIa	6 400	6 400	10 000	18 170
Plaice IV	169 000	169 000	175 000	13 656 †
Plaice VIIa	3 300	3 300	4 500	4 420
Plaice VIId,e	8 800	8 800	10 700	—
Plaice VIIf,g	1 700	1 700	1 900	1 450
Sole IV	27 000	27 000	27 000	1 010
Sole VIIa	1 300	1 300	1 500	805
Sole VIId	3 400	3 400	3 850	—
Sole VIIe	540	540	800	—
Sole VIIf,g	1 100	1 100	1 200	320
Sole VIIIa,b,c	4 700	4 700	5 300	—

* ACFM recommended a reduction in fishing effort of 30% ; TAC corresponds to reduction of 30% in rate of fishing mortality.
† TAC prior to any exchanges with Norway; minimum EC shares of the TACs are 91 000 t for cod, 76 910 t for haddock, 55 510 t for whiting and 13 656 t for plaice.

higher than those proposed by the Commission, whose proposals corresponded in most cases with the scientific advice (Table 4.3).

Summary

This system of TACs and quotas has proved considerably more robust than its founding fathers considered. In particular, the system has withstood not only the introduction of the *ad hoc* method for calculating quotas forced by the state of the North Sea haddock and cod stocks but also the modifications of the application of the Hague Preferences which followed, even though these did not have to be made. It was the fear that the conservation policy might collapse if controversial measures were proposed which led the Commission to continue to propose high TACs which did not represent any reduction in the high rates of exploitation. While it is easy to argue with hindsight, it would appear from what has happened that if the Commission had been more bold it might have been able to get the Council to adopt TACs which corresponded

Table 4.3 Recommendations of ACFM, the TACs proposed by the Commission and those adopted by the Council for 1992 for selected stocks for which analytical assessments were available, and the minimum TACs needed to meet Hague Preferences, where applicable, in tonnes.

Stock	Recommended by ACFM	Proposed by Commission	Fixed by Council	Minimum TACs to meet Hague Preferences
Cod IV	100 700*	98 000	100 700	110 723 #
Cod VI	12 000*	13 500	13 500	18 430
Cod VIIa	10 000	10 000	10 000	15 000
Haddock IV	61 000*	60 000	60 000	99 883 #
Haddock VI	10 800*	12 300	12 500	15 180
Whiting IV	120 000*	135 000	135 000	61 274 #
Whiting VI	6 500*	7 000	7 500	16 400
Whiting VIIa	9 700	9 700	10 000	18 170
Plaice IV	135 000†	170 000	175 000	13 656 #
Plaice VIIa	3 000*	3 000	3 800	4 420
Plaice VIId,e	7 820†*	9 600	9 600	—
Plaice VIIf,g	1 460	1 400	1 500	1 450
Sole IV	21 000†	21 000	25 000	1 010
Sole VIIa	1 240	1 240	1 350	805
Sole VIId	2 700†	2 700	3 500	—
Sole VIIe	770*	770	800	—
Sole VIIf,g	1 060	1 100	1 200	320
Sole VIIIa,b,c	5 000	5 000	5 300	—

* ACFM recommended a reduction in fishing effort of 30%; TAC corresponds to reduction of 30% in rate of fishing mortality.
† ACFM recommended 20% reduction in rate of fishing mortality.
†* VIId 800t + VIIe 2000t
# TAC prior to any exchanges with Norway; minimum EC shares of the TACs are 91 000 t for cod, 76 910 t for haddock, 55 510 t for whiting and 13 656 t for plaice.

with lower rates of fishing. Whether this would have resulted in a better situation today if it had been achieved is a matter of speculation. In general, catches have always exceeded TACs, it being the former, together with the size of year classes, which determines the evolution of the stocks, not the latter.

However, did the Commission ever have window of opportunity to propose lower TACs? It was not in a position to do so before the accession of Portugal and Spain and was then confronted by the problems posed by the accession of the two new Member States and the issues described. Member States are prepared to devote much time and energy to defending positions when settlement might be reached relatively quickly. This is for the reason that to settle quickly is regarded by their fishing industries as 'selling out'; the political pressure exerted therefore prevents quick settlements. This situation is not unique to fisheries. As described, both in the case of the application of the Hague Preference to North Sea haddock and that of 'flexibility' for western mackerel, the Commission did not present its proposals until the last possible moment, simply to avoid this situation.

It is not surprising that the level of TACs is mainly determined by political decisions because politicians regard it as their responsibility to respond to the pressures from their fishing industries as they consider fit. That is democracy in action. Account is taken of the scientific advice but more often than not it has been disregarded for socio-economic reasons, which is little more than coded language for saying 'avoiding political unpopularity'. Only when the consequence of disregarding the scientific advice would appear to be calamitous has it been acted upon, but often then not rigorously. It is not argued that decisions should be taken strictly in accordance with the scientific advice. However, it should form the basis of the conservation policy, enabling decisions to be taken which are consistent from year to year and which have the purpose of achieving specific objectives. This theme is developed fully in Chapter 12.

Chapter 5

Technical Conservation Measures and Control and Enforcement: Modifying the Regulations

or Complexity and Unenforceability

Technical conservation measures

Introduction

The history of technical conservation measures, as distinct from the conditions of fishing, has fallen into two phases. Each of these has concerned the implementation and modification of the two main regulations, Regulations No. 171/83 and No. 3094/86. The conditions of fishing, the technical conservation measures which form part of the annual TAC and quota regulations, also feature in this history but, with one exception, have not played a key political role in the development of this set of regulations.

The first regulation

Council Regulation (EEC) No. 171/83 was based upon the old NEAFC recommendations. As described in Chapter 3, it was not the first technical conservation measures regulation, Regulation No. 2527/80 having been adopted in September 1980, but it was the first of the agreed CFP. Its basic structure defined the different fishing regions and the standard minimum mesh sizes which applied in each of them. It then defined the species and regions in which it was permitted to use nets having minimum mesh sizes smaller than the standard, with special provisions when fishing for *Nephrops* and hake, followed by a series of provisions on permitted by-catches. It also defined minimum landing sizes as well as prohibitions on fishing for certain species in certain areas and limitations on the use of specific gears or vessels. The piecemeal format of the regulation reflected the fact that the NEAFC recommendations had been adopted at different times with no attempt being made to weld them into a coherent text. For example, the article on the use of small-meshed nets required reference to the subsequent article and three annexes; it was no wonder that the regulation proved difficult for both fishermen and enforcement officers to interpret and for the latter to enforce. In particular, the regulation was wide open to abuse because it repeatedly used the phrase 'when fishing for (a given species)' with no definition of what this meant.

However, for the same reason that the Commission did not propose major changes to the TACs, it did not attempt radically to modify this regulation immediately. Nonetheless, the regulation was modified six times in all. Many of these modifications had to be made because the original regulation contained review clauses concerning minimum mesh sizes, although all the reviews did was to maintain the mesh sizes in force, a subject described in more detail in Chapter 6. An important provision which was adopted was the specification of an area off the southwest of the UK and south and east of Ireland in which fishing for mackerel was prohibited. This provision included, for the first time, a definition of 'fishing for mackerel', a major triumph for the Commission which had to work hard to achieve this success. The sixth modification laid down the provisions required by the accession of Spain and Portugal. This series of modifications made a poorly drafted regulation even worse.

The second regulation

Minimum mesh sizes: the attempt at improvement

A major objective in drafting the proposal for the new regulation, which became Regulation No. 3094/86, was to avoid the expression 'fishing for (a given species)' by bringing together all the provisions which concerned area, minimum mesh size and by-catch limits. The manner in which this was finally achieved and which took many months of debate, was to include all these provisions in one annex to the regulation. A fisherman needs only to look at the annex to find the area in which he intends to fish, the main species he intends to catch and all the appropriate conditions, if any, which he has to respect, without any need for complex cross-referencing between different articles and annexes. The final table looked formidable because it included all permissible combinations of area, species, minimum mesh size, etc. Unfortunately, the Commission was unable to persuade the Council to simplify or delete any of the provisions specified in Regulation No. 171/83 so all the different exceptions appeared in its replacement.

From the practical point of view of fishing this did not matter; as few fishermen fish in more than two areas or fish for more than two groups of species, any individual fisherman is concerned with only a small section of the table. But it did nothing to simplify enforcement because the major problem in this area was, and still is, that, under Community law, fishermen are permitted to carry nets of as many different minimum mesh sizes as they wish. In consequence, it is almost impossible to prove which part of the catch has been taken with which mesh size and, therefore, whether the relevant conditions of fishing have been respected. True, catches have to be recorded in a logbook together with the area fished and the minimum mesh size used but no fisherman records information which could result in his prosecution.

The solution to this problem is easy to find. It is either to prohibit fishermen from carrying nets of more than one mesh size aboard or to specify that the catch composition must respect the minimum percentages of target species and maximum percentages of protected species provided in the regulation corresponding to the net on board having the smallest minimum mesh size, as is the case with present UK legislation. Proposals for such legislation were strongly resisted by the Member States because of their effects on the earnings of fishermen. It was argued that fishermen must have the right to switch between nets of different minimum mesh sizes appropriate to the species which they are fishing at different times of a trip. The Council refused even to consider adopting such legislation.

Neither was the Commission much more successful in getting the Council to adopt larger minimum mesh sizes. All that it did achieve was the adoption of an increase to 90 mm for the North Sea and the area west of Scotland north of latitude 56°00'N, but only to enter into force over two years after the regulation came into force.

Fishing in the continental coastal zone
The one area in which the Commission was successful in getting the Council to adopt measures which represented a significant improvement on Regulation No. 171/83 concerned fishing within the 12-mile coastal continental zone. This regulation made provisions which had the intention of banning fishing for sole and plaice with beam trawls by small vessels, defined as those not exceeding 70 GRT or 300 brake horse power (bhp). As explained in Chapter 2, the manner in which gross registered tonnage is measured permits the exclusion of certain enclosed spaces and fishermen were quick to utilize this loophole to construct large vessels which still corresponded with the limit. Brake horse power was to prove an equally flexible measurement; engines of the same cubic capacity can be adapted by means of turbo-chargers and carburettor modifications to deliver very different horse power and the makers were only too willing to provide engines 'derated' to a nominal power of 300 bhp which could quickly be modified to deliver far in excess of this. The result was the construction of a fleet of so-called 'Euro-cutters', specifically designed to take advantage of the regulations. Furthermore, Regulation No. 171/83 permitted these small vessels to use nets having a smaller minimum mesh size than the standard for the North Sea, 75 mm instead of 80 mm when fishing for sole for which it was required to have only 5% of sole on board, and this in the main nursery area for plaice and sole. Most of these Euro-cutters were Dutch and they were the bane of the Belgians and Germans. They bitterly resented Dutch fishermen using what had been introduced as a social, not as a conservation, measure, to pillage the coastal zone, especially as the tonnage limitation had been increased from 50 GRT, as it was in the NEAFC recommendation, to 70 GRT during the negotiations on the CFP.

Fortunately, the Dutch authorities also wanted to regulate the problem which had got out of hand. For once, there was a consensus of all the interested parties. As a consequence it was possible to adopt measures which almost put a lid on the problem. Beam trawlers permitted to operate in the zone had to be listed; those which had derated engines were permitted to be included on the list. Vessels could be replaced but were not permitted to have derated engines. The unenforceable measure of gross registered tonnage was dropped and engine power was defined, in accordance with modern definition, in kilowatts. However, because most of the German vessels operating in this zone conformed with the original tonnage and engine power specifications and could tow only light, long beams, special provision had to be made for these. For the same reason, Germany was not prepared to permit the minimum mesh size to be increased. However, it was a major step forward. Unfortunately, it was the only one.

Modifying Regulation No. 3094/86

The first to ninth modifications

Fishing for sole

Like its predecessor, Regulation No. 3094/86 was to be continually modified. Many of these modifications concerned beam trawling in the coastal zone and fishing for sole in the North Sea. In December 1989, a regulation providing that the standard minimum mesh size had to be used when fishing for sole in the first quarter of the year was adopted. This had the objective of preventing fishing on concentrations of sole which occur in deeper, warmer water when temperatures fall in the shallow inshore waters in cold winters; sole do not survive temperatures less than 2°C. Following the scientific advice, the area in which the restrictions on beam trawling for sole and plaice was extended in December 1988. At the same time, the maximum length of beam trawls was limited to 24 m. Faced with the fact that the power of every new beam trawler was greater than her predecessor, the Dutch authorities looked for a solution which would limit the ever-increasing fishing pressure on the stock of North Sea sole. They favoured a direct limitation on horsepower but given the problems associated with the enforcement of horsepower the Commission argued for a limitation on beam length, which the Council eventually adopted. This was linked with a new definition of beam length which was designed to overcome problems caused by the relentless ingenuity of fishermen in circumventing regulations. The original definition had been based on distance between the inside of the plates or skids which support the beam and which permit it to slide across the seabed. In order to use heavier gear fishermen simply extended the length of the beam and increased the width of the skids. Also, those who fished both within and outside the area to which restrictions applied had started to use beams whose length could be extended, thus creating the opportunity for further fraud. Maximum length was redefined as the total, overall length to which the beam could be extended. Unfortunately, in resolving one problem another was created. Fishermen fishing in the coastal zones where the maximum length of eight metres applied argued that the new definition reduced the effective length of their beams and their catching power by 20%. Despite the fact that they had no difficulty in catching their quotas and it could have been argued that this was a good conservation measure, they won the day and eight metres became nine metres.

Other modifications

Most of the other modifications to this regulation concerned technical changes. Many of these implemented changes in minimum mesh sizes when fishing for prawns and shrimps in the Skagerrak and Kattegat, which the Community had to adopt as a result of agreements with Norway and Sweden concerning conditions of fishing in these two areas, where management is shared. There were two measures which were particularly important. One was the adoption in 1987 of a measure to increase the minimum mesh size in the English Channel from 75 mm to 80 mm as from 1 January 1989 (an increase long recommended by the scientists and equally long proposed by the Commission). The other measure, equally long recommended and proposed and equally long opposed by the Member States was the extension of the eastward boundary of the 'mackerel box' by one degree to bring it in conformity with the scientific advice – 2°W instead of 3°W where the Council had previously determined it should run. This was finally adopted by the Council in October 1988.

Controversial proposals

The Proposal for the tenth modification
All these modifications had followed the normal pattern of events, being based on either scientific recommendations or agreements made with third countries. The Member States might have disagreed with the proposals of the Commission and refused to adopt them, or adopted them in a modified form, but at least there was no dispute about their basis. All this was to change with the adoption by the Commission of three proposals to modify Regulation No. 3094/86. The main contents of the first of these, adopted in July 1990, were:

- to increase the standard minimum mesh size in the North Sea and west of Scotland to 120 mm from the existing 90 mm, except for whiting;
- to specify that the codend should be constructed entirely of square-meshed netting (trawls are conventionally made of netting which, when stretched in the long direction of the trawl form a diamond; in square-meshed net, the netting forms a square when stretched in this direction);
- to introduce other specifications concerning the structure of the codend, such as the number of meshes in its circumference and their relation with the preceding piece of the trawl;
- to abolish the minimum landing size for whiting; and
- to prohibit the carriage of nets of more than one minimum mesh size, termed 'the one-net rule'.

The objective of proposing such a large increase in mesh size, together with the proposals on square mesh and other technical specifications concerning codends, was to reduce the high rate of discards, which were as high as 30% by numbers for North Sea haddock and whiting. (Discards are fish which are caught but then immediately returned to the sea because they are either less than the minimum legal landing size or are in excess of quota limits or there is no market for the fish.) Reducing the quantity of discards considerably was considered the most effective means by which to halt the rapid decline of the stocks of cod and haddock in the North Sea and west of Scotland which was occurring in 1990. These technical measures had the objective of ensuring that the meshes of the net stayed open. The proposals concerning whiting stemmed from the results of work on multi-species assessments which showed that whiting is a major predator and indicated that reducing the stock of whiting might result in higher catches of other species used for human consumption. The reason for the one-net rule has already been given.

Although there was no dispute about the state of the stocks concerned, there was no scientific recommendation to increase the minimum mesh size to 120 mm. No work at all had been carried out with codends made entirely of square mesh, only trials made with panels of square-meshed netting in a codend made of standard diamond-meshed netting. These trials had shown that these panels permitted small fish to escape more easily from the codends but also showed that 120 mm square-meshed net was equivalent to about 130 mm diamond mesh, which meant that the Commission was proposing an increase in minimum mesh size of effectively 40 mm.

The Commission did not help its case by not consulting its own Scientific and Technical Committee for Fisheries until after it had adopted the proposal and then at short notice. The STCF was politely scathing. It commented that, although discards would be almost completely eliminated by increasing minimum mesh size to 120 mm, there would be long-term losses of 6% for cod, 34% for haddock and 30% for whiting, even based upon the multispecies assessment. Concerning whiting, the STCF noted that the result would be to increase the size of the stock, not to reduce it, thus increasing predation on other species. It also stated that 'preliminary examination at national level of the historical proportion of catches comprised of whiting has not revealed any fleet which appears capable of operating within the derogation'. In simpler language, the STCF did not consider that any fisherman would be able to comply with the provision that the catch must consist of at least 50% by weight of whiting and no more than 10% by weight of protected species which it was proposed should apply to using nets having a minimum mesh size of 90 mm when fishing for whiting. Commenting upon the technical specifications for codends STCF stated that: 'The design specified . . . is completely novel and therefore the selectivity and practical utility of this gear is unknown'.

If the STCF had to be circumspect in expressing its views the fishing industry did not. It reacted furiously. The proposed increase in minimum mesh size affected those fishermen fishing mainly for cod, haddock and whiting, in particular the Scots, but also British fishermen in general. Up until this point a proposal to increase a minimum mesh size by 10 mm had always been regarded as economically disastrous by the industry and the step eventually adopted had invariably been 5 mm. For the Commission to propose an increase of a nominal 30 mm, which was more likely to be effectively 40 mm, was unbelievable. They argued that the short-term effects of the proposals, if adopted, at a time when catches were already falling drastically would be to bankrupt them. Even if they survived, the long-term effects would be unacceptably serious. For obvious reasons, this group of fishermen did not believe the results from the multi-species model. It was Denmark which was convinced of the results from this model but even for Denmark the proposed minimum mesh size of 90 mm was unsatisfactory; it wanted 70 mm. France also opposed the proposal, wanting a mesh size smaller than 90 mm for whiting, at least in the southern North Sea but for a different reason than Denmark.

French fishermen fished whiting for human consumption and had always considered 90 mm too large a mesh size for whiting.

No fishermen were satisfied with the technical measures concerning codends. As the STCF had noted, they were an unknown quantity. There was considerable enthusiasm amongst many fishermen fishing for species such as cod, haddock and whiting, collectively termed 'roundfish' for the introduction of square mesh panels which had been shown to facilitate the escape of these species. In contrast, those fishermen fishing for sole and plaice ('flatfish'), for which square mesh did not appear to be effective and might reduce escapement, were opposed. There was general opposition to the proposed one-net rule. The Commission was unsympathetic. The state of the stocks with which the proposal was concerned was rapidly approaching a disastrous state according to the scientific evidence and if measures were not adopted to arrest the decline, their collapse was on the cards in which case no-one would be be able to fish.

The proposal for the eleventh modification
Instead of revising its proposal in the light of all these criticisms, as might have been expected, the Commission adopted another proposal in December which, in some respects, went even further down the same path as the previous one.

The main items in this proposal were:

* to ban the use of driftnets longer than 2.5 km;
* to ban beam trawling in Region 3 (Bay of Biscay);
* to increase the standard minimum mesh size in Region 3 from 65 mm to 80 mm and that for fishing for *Nephrops* from 55 mm to 65 mm;
* to introduce seasonal prohibitions on fishing with trawls and Danish seines in areas off the coasts of Spain and Portugal; and
* to limit fishing for sole in the North Sea with nets of 80 mm to south of latitude 55°N.

The proposal to increase minimum mesh sizes in Region 3 was a straightforward conservation measure having as its objective reducing the enormous catches of small hake caught in that area. The seasonal bans on fishing, which had the same objective, already formed part of the TAC and quota regulation for 1991 so this represented only their logical transfer to this regulation. The proposal on North Sea sole was another step in tidying up the regulations concerning this fishery; it, too, was a transfer from the TAC and quota regulation. The other two proposals had political overtones.

The proposed ban on the use of driftnets stemmed from the adoption by the General Assembly of the United Nations of a resolution prohibiting the use.of large-scale pelagic driftnets, for which the Community had voted. The Commission had no option but to propose a measure implementing this ban, choosing 2.5 km as an arbitrary definition of 'large-scale'. Driftnetting is widely practised in all Community waters so to ban all driftnetting would have been politically impossible. The proposed ban affected only fishing for albacore tuna, which had been slowly developing in the Community, was not yet of major importance and which was the only fishery in which the by-catch of marine mammals was a political issue. It also had the beneficial effect of resolving an internal dispute between France, which had developed this fishery, and Spain, which was violently opposed to it, essentially for social reasons. Spanish fishermen traditionally fish for tuna using pole and line, a manpower-intensive method of fishing compared with driftnets. The Spanish feared that they would be forced for economic reasons to switch to driftnetting, increasing unemployment in an area where this was already high. As most of the fishermen lived in the Basque region, this had potentially very serious implications for the Spanish authorities. Italy and the UK supported Spain.

The reason given in the Explanatory Memorandum proposal to ban beam trawling was as follows:

'Fishing with beam trawls has an impact on the soft sea-bed in this region, which poses a threat to benthic ecosystems. The effects of beam trawling on catches of commercial species over this type of sea-bed have not been assessed. Therefore, in view of the lack of evidence, the use of this gear should be prohibited.'

On the face of it this was a case, the first, of the Commission applying the precautionary principle to fisheries. This principle is that of taking action to avoid adverse effects which it is considered might occur before proof has been obtained that the action causes the effects. In actual fact, the proposed ban resulted from French pressure on the Commission. France has few beam trawlers and none in the Bay of Biscay area. Over the years there had been several incidents between Belgian fishermen, who have the right to fish for sole in the Bay of Biscay, using beam trawlers, and French fishermen who objected to what they considered the adverse effects of beam trawling on their fisheries for *Nephrops*. Legally the Commission was skating on thin ice because Regulation No. 171/83 provides that:

'The conservation measures necessary to achieve the aims (of the conservation policy) shall be formulated in the light of the available scientific advice . . . '

of which there was none to support the proposed ban.

The proposal for the twelfth modification
The Commission produced yet another proposal to modify Regulation No. 3094/86 in June 1991 whose main points were:

- to clarify the method by which crustacean and molluscs should be measured;
- to transfer to this regulation from the annual TAC regulation provisions concerning fishing for sprat and herring;
- to ban the use of automatic sorting equipment on vessels larger than 26 m, with certain exceptions;
- to rationalize the annex laying down minimum mesh sizes and the conditions attached to their use; and

- to increase minimum landing sizes for mackerel, Spanish mackerel, anchovy and scallop and to introduce them for mackerel, Spanish mackerel, anchovy and crab in areas where they did not apply.

Regulation No. 3094/86 provides two and sometimes three measurements which may be taken for determining the minimum landing sizes of crustaceans and molluscs, the reason being that often only parts of these animals are landed. Also, some methods are better than others; for example, carapace (the hard headshell) length of crustaceans is better than total length because the former is inflexible whereas the tail can be stretched. The original wording of the regulation stated that '. . . crustacean or mollusc is undersized if it is smaller than any of the minimum sizes specified' – that is, it was undersized if any of the measurements were less than the minimum landing size. The proposed new wording reversed this situation; the animal needed to be larger than only one of the provided measurements. The proposed rewording also deleted the words 'either' and 'or' which preceded the definitions, implying that only one needed to be measured. Although the proposed rewording certainly clarified the text, it represented a step backwards in conservation terms, which was curious in a proposal which was being presented by the Commission as a major advance.

The proposal to transfer the provisions concerning seasonal bans for fishing for herring and sprat off the west coast of mainland Denmark was a logical step; these provisions had appeared unchanged in the annual TAC regulation since 1983 and, as a permanent measure, should be part of the regulation on technical conservation measures.

The objective of the increase in minimum landing sizes and the ban on sorting machines was to reduce discards by reinforcing the minimum mesh size regulations and reducing the economic incentive to catch large quantities of small fish. However, the Explanatory Memorandum linked this with species for which a high proportion were likely to survive or those caught by highly selective methods of fishing, neither of which is the case for mackerel, Spanish mackerel and anchovy. The proposed ban on sorting machines was much more logical and one which had been proposed in the early days of the conservation policy and rejected.

The main changes proposed for minimum mesh sizes were to increase that for the Irish Sea from 70 mm to 80 mm – another move which had been long sought by the Commission but equally long rejected by the Council – that for herring from 32 mm to 40 mm and for mackerel from 16 mm (32 mm in the North Sea) to 65 mm. It was also proposed to increase the minimum mesh size for horse mackerel from 16 mm to 65 mm and for blue whiting and argentines from 16 mm to 40 mm and for most species of prawns from 30 mm to 40 mm. All these proposals excluded the Skagerrak and Kattegat for the reason that agreement was needed with Norway and Sweden for changes in the minimum mesh sizes which applied there.

The debate and its conclusions

These three very controversial proposals produced very heated reactions from the fishing industry, especially in the UK which would have been the most affected if they had been adopted as they stood. The SFF and the NFFO described them as 'devastating'. The former argued for retaining 90 mm diamond mesh but with 80 mm square-meshed panels while the latter, whose fishermen caught a higher percentage of cod and plaice, considered that 100 mm was acceptable. The SFF backed up its case by undertaking a two-week trial with 120 mm square-meshed netting during which it caught eight haddock and no whiting of legal size but rather spoilt its case for retaining 90 mm diamond-meshed netting by catching 35% undersized haddock and 38% undersized whiting with that gear used as a comparison. Danish and German fishermen, whose catches consisted predominantly of cod, saithe and plaice found the proposal for 120 mm acceptable, if not the technical details concerning codends. Those Dutch fishermen who fished for cod and plaice were also in favour of the proposal while those who fished for sole were not affected, being able to continue to use 80 mm. As already stated, Danish and French fishermen opposed the measures concerning whiting but for different reasons.

Vice-President Marin defended his proposals vigorously as being 'an extreme solution to an emergency situation' but his argument was undermined by the former chairman of ACFM stating that 120 mm was suitable only for a fishery in which catches of cod predominated.

The Council (Fisheries) met on 20 October 1990 but was unable to agree, except not to agree to 120 mm. It was hinted that the final compromise might be either 100 mm diamond mesh with a 90 mm square-meshed panel (100 mm/90 mm) or 110 mm/100 mm. The SFF stuck to its position of 90 mm/80 mm; feelings ran so high in Scotland that a fisherman doused the Director-General for Fisheries, Mr Almeida Serra, with a bucket of water when he visited Peterhead.

South of the border, the NFFO was willing to accept 100 mm/90 mm with a derogation for fishing whiting with 90 mm in the winter. In France and in the south-west of England, fishermen engaged in driftnetting for albacore tuna fought against the ban on large driftnets.

The Council (Fisheries) met again on 8 July 1991 to re-examine the proposals and failed yet again to agree, leading Vice-President Marin to threaten that he would let Member States adopt their own conservation measures if there was no agreement at the next meeting of the Council, scheduled for October. Whether this was an empty threat or not will never be known because agreement was reached at that meeting. However, for reasons which have never been explained, the regulation was not adopted until 27 January 1992.

The main measures adopted for Region 2 were:

- an increase in the standard minimum mesh size to 100 mm with the option to use a panel of 90 mm square-meshed netting;
- a limit of a maximum of 100 meshes in the circumference of codends having a mesh size equal to or greater than 90 mm;

- maintenance of the minimum mesh size for whiting at 90 mm on condition that catches consisted of 50% whiting and included no more than 10% of cod, haddock and saithe, which was subsequently modified to 70%;
- modification of the minimum mesh sizes, in millimetres, for Regions 1 and 2 (the sea areas in the north-east Atlantic north of latitude 48°N), except the Skagerrak and Kattegat, as follows:

Species	Proposed	Adopted	Previous
Herring	40	40	32
Mackerel	65	40	16*
Horse mackerel	65	40	16
Blue whiting	40	40	16
Argentines	30	30	16

*32 mm in the North Sea

- reduction of the minimum landing size for whiting from 27 cm to 23 cm;
- a ban on driftnets longer than 2.5 km, with a two-year exemption for those fishermen who had an historic record of using this gear who would be permitted to use nets of up to 5 km for two years unless it could be scientifically proved in the meantime that there was an 'absence of ecological risk linked thereto';
- a ban on the use of sorting machines on vessels fishing for mackerel, herring and horse mackerel except for freezer vessels which can carry them for grading;
- transfer of the measures laying down seasonal bans on fishing for herring and sprat off the west coast of mainland Denmark to the regulation;
- adoption of the new provisions concerning measuring crustaceans and molluscs;
- generalization of the minimum landing size for mackerel to 20 cm, compared with 30 cm proposed;
- deletion of the derogation permitting the use of nets having a mesh size of 75 mm by vessels of less than 221 kW engine power fishing for sole in the North Sea;
- adoption of 80 mm minimum mesh size for the Irish Sea.

In contrast to the measures adopted for Region 2, none of the proposals made for Region 3 (waters south of 48°N) were adopted, except for a minimum landing size for anchovy and the seasonal bans on fishing in specified coastal waters of Spain and Portugal. The latter represented only the advance of making the measures permanent since they already existed in the TAC and quota regulation.

Many of the measures, such as the increase in minimum mesh sizes for pelagic species, the deletion of the derogation for sole and the adoption of 80 mm minimum mesh size in the Irish Sea, represented small but significant advances. But whether the increase in the minimum mesh size to 100 mm with the restrictions on the number of meshes in the circumference of the codend will achieve its objective of reducing discards remains to be seen because such measures are easy to circumvent and difficult to enforce. Some of the measures appear likely to have adverse effects. The generalization of the minimum land-ing size for mackerel may increase discards instead of decreasing them, as is its objective, for the reason that mackerel are caught with purse seines which are non-selective for size, as well as pelagic trawls. Any small fish caught will have to be discarded instead of landed, as they could be previously, unless the ban on carrying sorting machines stops fishermen fishing in areas where small mackerel are abundant. Alternatively, the fish will be landed and graded ashore, small mackerel being sent for reduction to fish meal. The new provi-sions on measuring crustaceans and molluscs will certainly allow the landing of smaller crustaceans, particularly lobsters, than hitherto because it will be possible always to use total body length, which can easily be stretched.

The most controversial measures adopted were those on whiting. These pleased neither Denmark nor the UK. All available evidence indicates that it will be impossible to use a 90 mm mesh size for whiting and take catches which consist of at least 50% whiting and no more than 10% of cod, haddock and saithe. The Danish authorities, with financial assistance from the Com-mission, are conducting trials to show that it is possible. Certainly Denmark wanted the mesh size to be 70 mm, as probably did France but for the differ-ent reasons described. It would have seemed more logical, if the Council considered that the stock of whiting should be reduced, to have permitted the use of nets having a mesh size of 16 mm and to have deleted the minimum landing size. As it is, the measure appears to fall between two stools, satisfy-ing neither those fishermen who consider whiting a pest to be destroyed and those who consider it a valuable species to be caught and sold for human con-sumption. The regulations as adopted appear to be the result of a classic com-promise (see Chapter 1).

As far as Region 3 is concerned, except for the measures to protect the nurs-ery grounds for hake, there was almost no improvement in the technical con-servation measures. Once again, France, Spain and Portugal, with almost certainly France in the van, were able to block any increases in minimum mesh sizes in Region 3. French fishermen having an historic record were even allowed to continue to use large pelagic driftnets for at least two years.

The history of the implementation of technical conservation measures has been that it is impossible to get the Council to advance rapidly or beyond certain limits. In this case the Commission used the rapidly declining size of the stocks of North Sea and west of Scotland stocks of haddock and whiting to panic the Council into adopting a very large increase in the minimum mesh size. Even in these circumstances it did not succeed although it must be admit-ted that the increase of 10 mm in one step was more than had ever been achieved previously. However, over a year was lost in furious debate. Given that urgent action was necessary, it is difficult to understand why the contro-versial measures on whiting were proposed at the same time, which were obvi-ously going to make the issues more difficult to resolve. The measures which were adopted formed the eleventh modification.

The tenth modification concerned the ban on fishing for mackerel in the area to the south of the UK and south and east of Ireland. This would have expired on 1 January 1992 if the Council had not made a decision by

30 November 1990. Instead of fixing a new date for review, as might have been expected, the Council made the ban permanent. At one stage it had appeared as if this proposal would have given rise to controversy because Scottish fishermen have for years wanted the ban modified or repealed but in the event it slipped though the Council.

The twelfth to fifteenth modifications

Regulation No. 3094/86 has continued to be modified, despite the rules concerning codification. Several of the modifications provided derogations from the legislation in the eleventh modification. The twelfth modification permitted the continued use in Regions 1 and 2 until the end of 1992 of a minimum mesh size of 32 mm for mackerel, horse mackerel, herring, pelagic cephalopods, pilchards and blue whiting and 30 mm for prawns. The reason given in the preamble was that fishermen had not had sufficient time to obtain nets of the new mesh sizes. The derogation concerning the minimum mesh size of 32 mm was extended to 31 December 1993. This was done not by an amendment of Regulation No. 3094/86 but by an article in the TAC and quota regulation for 1983 (Regulation No. 3919/92). The reason given in the preamble to the regulation in this case was that '. . . new studies are at present in progress to evaluate the effects on selectivity on individuals enmeshed whilst fishing for these pelagic species . . .' and that '. . . these new elements may contribute to a redefinition of this type of fishing'. (It is assumed that the individuals referred to meant not fishermen but fish whose heads become stuck in the meshes of the larger mesh size.) This effectively became the fifteenth modification, although legally it did not count as such. It is unusual for the Council to modify one regulation by provisions made in another. Legally it is a very unsatisfactory procedure. Practically, it makes it difficult for those who have to use the legislation to be aware of the modification.

The twelfth modification also allowed vessels whose engine power was 221 kW or less to continue until 31 December 1992 to use nets having a minimum mesh size of 75 mm when fishing in the North Sea south of latitude 55°N.

The thirteenth modification added several provisions which were to apply until 31 December 1992 concerning fishing for whiting in the North Sea and west of Scotland north of latitude 56°N. These were that (a) all catches were to be treated as if they were taken with the net having the smallest minimum mesh size on board, (b) it was prohibited to carry on board a net having a minimum mesh size less than 90 mm, (c) cod, saithe and haddock should not be more than 10% of the protected species on board and (d) plaice should not be more than 10% of the weight of protected species on board. The percentage of whiting was to be calculated as a proportion of the combined catches of whiting, haddock, saithe and cod on board. The provisions were not made permanent for the reason given in the preamble that '. . . scientific analyses should nevertheless be completed and the adoption of a permanent redefinition of the conditions under which this fishery may be conducted would therefore be premature'.

The fourteenth modification '. . . prohibited to undertake encirclements with purse seines on schools or groups of marine mammals when aiming to catch tuna or other species of fish'. The regulation applied the provision to all Community vessels irrespective of where they were fishing. This provision essentially concerns catching dolphins when purse seining for tuna, a problem specific to eastern Pacific waters. A few Community vessels do fish there for tuna but the legislation is essentially cosmetic because it cannot be enforced .

The TAC and quota regulation for 1983 also modified the bycatch limits when fishing for Norway pout. The maximum bycatch was increased to 15%, of which no more than 5% could be cod and haddock combined. It is incredible that this modification was adopted without any comment in the British fishing press. In 1984 the Commission had proposed a similar modification which was bitterly opposed. Scottish fishermen carried out a long campaign which included several fishing trips to show that what the Commission was proposing was impossible to achieve in practice. A regulation, valid for eight months, was eventually adopted but no attempt to modify these provisions was ever made again until 1993.

Conditions of fishing

Seasonal bans on fishing

As explained previously, the conditions of fishing are the technical conservation measures which appear in the annual TAC and quota regulation. The reason for separating them is that conditions of fishing are measures which are supposedly temporary and need to be reviewed each year to see whether they need to be renewed whereas technical conservation measures are permanent. Each TAC and quota regulation expires at the end of each year to which it applies so the conditions of fishing therein automatically end on the expiry of the regulation. In contrast, the technical conservation measures regulation is of indefinite duration. In practice, most of the conditions of fishing have been adopted unchanged, year after year, with no debate. Even in those cases in which they have been modified, there has been little political controversy, probably because the debate on TACs has taken centre stage. For example, the seasonal bans on fishing for herring and sprat off the west coast of mainland Denmark were first adopted in the TAC and quota regulation for 1983 and have remained unchanged ever since. As described in the previous section, they have now been transferred to the technical conservation measures regulation. Similarly, the seasonal bans on fishing off the coasts of Spain and Portugal have been transferred, in this case only one year after they were first adopted. Also, to the extent that they have been reviewed, the new measures have made significant advances over those previously in force. For example, fishing with nets having a minimum mesh size of 16 mm in the Skagerrak and Kattegat, which was originally permitted throughout most of the year by all vessels, was gradually phased out by 1988 for vessels equal to or longer than 25 m and by

1990 for smaller vessels. This measure had the objective of stopping vessels fishing for juvenile herring under the guise of fishing for sprat. It hit the local fishermen hard because they had been accustomed for fishing for small clupeoids (herring-type fish) for many years, long before there were any regulations. When sprat was the dominant species this presented no conservation problem but when herring replaced sprat, there was an outcry about the fishery from those countries which fished adult herring. Denmark, the only Community country concerned, had little choice in accepting the more stringent measures; it has only three votes and all other Member States are opposed to industrial fishing. But by not making a political issue of the matter Denmark was able to obtain a slow, rather than a rapid phaseout. However, the situation is not as clear-cut as it appears because a TAC for sprat is fixed for this area and the TAC and quota regulation includes a provision which makes it possible for the Community share to consist entirely of herring. The only measures to provoke real controversy have been those to limit the number of days at sea.

Limitation on days at sea

This controversy had its roots in a proposal of the Commission that, for 1990, all fishing by a Member State in the North Sea should stop once it had exhausted its quotas of cod, haddock, whiting and saithe combined. The objective was to prevent fishing continuing for other species, in which case the catches of species for which the quotas are exhausted have to be discarded because it is legally prohibited to land them. The proposed measure seemed to the Commission a small step towards better conservation but it was rejected by the Member States which argued that Article 5 of the basic conservation regulation made them responsible for managing their quotas. However, it resulted in the Commission forcing the UK to adopt national legislation limiting the number of days at sea by those of its vessels which traditionally fished for cod and haddock.

In the light of the ACFM advice for 1991 that fishing effort on the stocks of North Sea and west of Scotland cod, haddock and whiting should be reduced by 30%, the Commission proposed that all vessels which historically caught a high proportion of these cod and haddock stocks (the stocks of whiting were not threatened and had been included by ACFM in its advice only because they are caught in a mixed fishery with cod and haddock) should have to spend ten consecutive days in port each calendar month. Notwithstanding the furious reaction of the fishing industry, mainly the UK and in particular the Scots who were most affected, the proposal was adopted in the form of a consecutive eight day tie-up in each calendar month for the period 1 February to 31 December.

For 1992 the Commission proposed an even more severe measure, a tie-up of 200 days a year. This was greeted with consternation by the industry which, after intensive lobbying, managed to achieve a reduction to either 135 days or eight consecutive days a month in port for the period 1 February to 31 December. For 1992 similar limitations applied to vessels fishing for cod in the Skagerrak and Kattegat.

Provision was made in the regulation that any Member State which could not catch its quotas as a result of the application of these rules could apply to the Commission for exemption. Despite there being no other provision, the Commission totally exempted from the tie-up vessels which fished throughout 1991 with nets having a minimum mesh size of 110 mm. For 1992 it made similar provisions; vessels using a minimum mesh size of 100 mm had to tie up for only 67 days while those using 120 mm or larger were totally exempted. Not only did this create a peculiar legal situation but, more importantly, had serious implications for the effectiveness of the conservation measure, as further discussed in Chapter 6.

Control and enforcement

As described in Chapter 3, the first control regulation adopted, Regulation No. 2057/82 establishing certain control measures for fishing activities by vessels of the Member States, laid down a series of administrative provisions. Between 1982 and 1987 these were altered in only two significant respects. First, the provisions applying to trans-shipment were strengthened, almost entirely at the urging of the UK which was experiencing severe problems in effectively controlling the large number of trans-shipments taking place in its waters. Second, provisions were adopted whereby any Member State which had been unable to fish all of its quota in a stock because overfishing by another Member State had resulted in fishing on that stock being stopped could claim compensation. This consisted of a forced transfer of quotas from the transgressor to the prejudiced Member State, the quantities by stocks being proposed by the Commission. This was settled in the Management Committee unless the Member States concerned settled the matter previously by an appropriate exchange of quotas. This latter measure proved to be the one provision of the control regulation which has worked satisfactorily.

In 1987 Regulation No. 2057/82 was codified, being replaced by Regulation No. 2241/87. Codification is a purely technical process which does not involve any changes whatsoever to the provisions of regulation being codified; the Commission does not make a proposal and there is no possibility therefore to amend the provisions. Again at the instigation of the UK this regulation was considerably strengthened by a modification adopted in 1988 to control the activities of vessels flying the flag of or registered in one Member State which consistently landed in another. In the case of the UK it was UK-registered vessels which were crewed by Spanish fishermen and operated from Spanish ports, rarely or ever entering UK ports, the so-called 'quota-hoppers'. The UK had been infuriated by the fact that the landings of these vessels in Spain in 1987 had been reported late to the UK with the result that the UK's quotas of several stocks had been overfished, which led it into trouble with the Commission. The outcome was the adoption of a series of measures which placed the onus of reporting catches and enforcing the national legislation of the flag state on the Member State in which the vessel landed, against whose quotas the landings could be debited if it failed to do so. Another provision intended to make enforcement of conservation measures was that adopted; a 'quota-hopper' with a bad record had to have a certificate from the flag state stating that it had been inspected by the flag state within the last two months otherwise it was banned from landing outside the flag state.

However, any legislation is only as good as the extent to which it is enforced and, as described in Chapter 7, enforcement was to prove the Achilles' heel of the conservation policy.

Chapter 6

Political Expediency *v.* Scientific Advice

or *Avoiding Taking Decisions*

The background

The legal basis

Article 2 of Regulation No. 170/83 provides that the measures necessary to achieve the objectives of the conservation policy '. . . shall be formulated in the light of the available scientific advice . . .'. This chapter examines, by means of a series of case histories concerning minimum mesh sizes, seasonal bans on fishing and the fixing of TACs, how political factors affect the implementation of the scientific advice. This creates a vicious circle because the political considerations inevitably influence the manner in which the advice is given which in turn influences the manner in which it is implemented.

The problems in presenting scientific advice

Fisheries scientists are faced with two major problems in presenting their advice which have repercussions on the way in which it is perceived and acted upon, these being (1) the lack of specific management objectives and (2) the manner in which to present their advice.

As long ago as 1981 ACFM stated that it should be the role of 'managers' (meaning politicians and fisheries administrators) to determine objectives and management strategies. ACFM considered that its proper role as a provider of advice on fishery management should be only to 'evaluate the biological consequences of these management strategies or define the biological constraints for the attainment of these objectives'. ACFM stated that if managers failed to provide objectives and strategies it would provide advice but that this would be based upon biological considerations, the only objectives it was competent to determine. Because the Community does not have specific management objectives, a subject discussed in detail in Chapter 11, the Commission is unable to request specific advice from ICES, except very occasionally, and has perforce been able only to present very generalized requests concerning the state of the stocks and short-term catch predictions. (ICES is the organization which provides the advice. ACFM is the committee established by ICES to formulate that advice.)

Identical comments apply to NEAFC which requests ICES for advice on the same stocks as the Community. In consequence, the advice given by ACFM is based upon biological considerations. ACFM has placed greatest emphasis on two considerations. One of these has been the need to reduce rates of fishing mortality to those corresponding with one or other of the biological reference points used by ACFM, notably F_{max} and $F_{0.1}$, in order to maximize long-term yields. The other has been to ensure that the sizes of spawning stock biomasses do not fall below the target levels considered the minima by ACFM to ensure that recruitment failures do not occur. 'Managers' have been only too willing to abdicate their responsibilities in this sphere because it enables them to reject the scientific advice when it suits them but to blame the scientists when they take actions which are unpopular.

The presentation of the advice plays a crucial role in the manner in which it is interpreted and implemented. Fisheries science is admittedly an inexact science and advice must, therefore, reflect its uncertainties if the advice is to retain its integrity. On the other hand, both politicians and administrators are notorious for demanding concise summaries on which to base their actions. Prior to 1987, the reports of ACFM were written in a narrative style and contained much technical detail about the problems associated with making the assessments on which the advice was based. Most, if not all, administrators probably considered themselves far too busy to read the reports. Few, if any, had the necessary background knowledge to understand them fully, if at all. The administrators depended upon their scientific advisors to present them with a summary. Recognizing this situation, STCF decided to short-circuit this step and present its reports in a standard, simple format which it hoped administrators would read and were capable of understanding. From 1987 onwards, ACFM also adopted this format, in which information is presented under eight headings with a brief commentary or, where the state of the stock justifies it, a longer overview of the main issues. All graphs and tables are presented as appendices.

Most administrators now consider that they can understand the reports without scientific assistance but it is almost certain that this method of presentation has had adverse repercussions. Administrators are primarily interested in deciding at what level to fix TACs. That part of the format which provides this information is a table showing selected catch options for a given year (for example, 1993) and the corresponding consequences for the spawning stock biomass at the start of the following year (1 January 1994). Because ACFM places so much emphasis upon maintaining minimum levels of spawning stock biomasses, the Council now frequently bases its decision about the appropriate level of a TAC only upon the criterion of whether the spawning stock biomass will be kept above the minimum level determined by ACFM. This process of decision taking ignores several considerations. First, it ignores the fact that the advice is short-term, for one year only, and takes no account of the long-term implications for the spawning stock biomass. Second, it ignores other consequences, such as that the TAC fixed on this basis might increase an already excessive rate of fishing upon the stock leading in the long term to lower catches, smaller spawning stock biomasses and poorer economic returns. Third, and most importantly, it ignores the fact that the advice of ACFM in this respect is purely precautionary. It has never been established for any stock of fish in Community waters that having more spawning fish results in having larger numbers of young fish, a subject examined in more detail in Chapter 8. Unfortunately, most administrators and all fishermen consider it self-evident that having more spawning fish must result in having more young fish and so the advice of ACFM corresponds with their intuitive, but unfortunately incorrect, ideas about how fish stocks should be managed. What is so ironic about the situation is that the wealth of information provided in the appendices of the report of ACFM is largely, if not totally ignored. It is questionable whether any administrators refer to the graphs or, even if they do, understand their implications. It is commonplace for administrators to consider that they are constrained to choosing from the selection of options provided in the simplified format, whereas the graphs provided an infinite number of options.

Regrettably it must be concluded that, although the scientific advice is of high quality, the majority of those who use it do not have the capability of taking full advantage of the scientific advice. Certainly, the latest presentation of the advice has not improved the understanding of the advice with deleterious consequences for its implementation.

The role of the Commission

The fishing industry frequently complains that it is subject to arbitrary decisions which are based only upon scientific advice and that the Commission disregards all other considerations. This is to ignore the fact that Article 2 of Regulation No. 170/83 provided for conservation measures to be based upon the available scientific advice. The provisions of Article 4 of Regulation No. 3760/92, which has replaced Regulation No. 170/83 are far broader, providing that '. . . (conservation) measures shall be drawn up in the light of the available biological, socio-economic and technical analyses . . .'. However, until the end of 1992 the Commission had to act in accordance with Regulation No. 170/83 and was constrained by its Article 2 to base its proposals on the scientific advice. When it did not do so, the Council used it as a reason for not adopting what the Commission proposed. But there was, and still is, an additional reason. The advice of ACFM and STCF is the only advice available. No other organization produces internationally accepted reports which cover other aspects of fisheries, such as economic assessments. Furthermore, while individual scientists may disagree with the interpretation placed upon the scientific data, the advice of ACFM is usually unanimous. The advice of ACFM was the bedrock on which the Commission could make proposals and argue its case.

The role of the Council and the fishing industry

It is the Council which decides, albeit on a proposal from the Commission, and the Council is under no such constraints as the Commission. Regulation No. 170/83 provided that the objectives of the conservation policy were, *inter alia*, the '. . . balanced exploitation of (the biological resources of the sea) on a lasting basis and in appropriate economic and social conditions . . .'. This provision permitted each Member State to use whatever argument, scientific, economic or social, which best supported its case at the time, permitting the Council to adopt less severe measures than those proposed by the Commission or to postpone taking a decision or take no decision whatsoever. Because the scientific advice was not given in response to objectives laid down by the Council, it felt free to ignore advice based solely on biological objectives. However, if the Council considered that it was left with no option but to take drastic action in accordance with the advice, the Council invariably sought to absolve itself from all responsibility, blaming the scientific advice for 'leaving it no alternative', even though the situation might have resulted from the past inaction of the Council in ignoring the scientific advice over many years. The usual situation provoking such action was when ACFM advised that a stock was in imminent danger of collapse. This state of affairs is unlikely to change with the replacement of Regulation No. 170/83 by Regulation No. 3760/92.

But the Council is the ministers meeting together and they present the interests of and advance the arguments provided by their fishing industries to justify the decisions which they take. Thus the decisions taken by the Council are largely influenced by the fishing industry, even though it often objects to the results because every decision usually represents a compromise. Because of the influence of the fishing industry upon the decisions of the Council, it is impossible to consider their actions separately in analyzing their influence on what decisions were taken and how the proposals of the Commission were modified.

Minimum mesh sizes

The history of increasing minimum mesh sizes in the North Sea is described in detail because it illustrates all the arguments which are used to delay increases. The other examples are provided to show that what happened in the North Sea was not an isolated case.

The North Sea

The history of minimum mesh sizes
The history of minimum mesh sizes in the North Sea predates the Community by more than two decades, ICES having recommended a minimum mesh size of 75 mm for the North Sea as long ago as 1934. This was implemented by a few nations in 1937 but the outbreak of the 1939–45 war resulted in its abandonment. After the war, in 1946, the UK government convened an 'Overfishing Conference' which recommended a minimum mesh size of 80 mm for waters south of latitude 62°N and north of latitude 48°N and between longitudes 32°E and 42°W, excluding the Baltic Sea, an area which now includes all Community waters. By the time that the Convention was ratified in April 1954, the minimum mesh size had been reduced to 75 mm, but 80 mm was to be introduced as from April 1956.

With responsibility for fisheries in the hands of the Community, the

Commission as part of its proposal for a regulation on technical conservation measures proposed in 1976 a minimum mesh size of 90 mm for the North Sea and eastern English Channel. As described in Chapter 3, the Council did not adopt this proposal. However, by Regulation No. 2527/80 the Council fixed the minimum mesh size at 70 mm for codends made of single twine and 75 mm for those made of double twine. This was a step backwards because most of the countries which formed the Community in 1980 had been signatories to the Overfishing Convention which had agreed to 80 mm to enter into force in 1956. But Regulation No. 2527/80 did provide that, as from 1 December 1980, the minimum mesh size was to be increased to 80 mm followed by a further increase to 90 mm with effect from 1 October 1982. But all these provisions lapsed when Regulation No. 2527/80 was not prolonged in November 1981 when Denmark invoked the 'Luxembourg compromise'. Regulation No. 171/83, adopted on 25 January 1983, fixed the minimum mesh size at 80 mm with immediate effect but deferred the introduction of 90 mm by 15 months, to 1 January 1984. But the ink was scarcely dry on this legislation before the Council, by unanimous decision and not on a proposal from the Commission, delayed its implementation by a further year, subject to review. This was the one and only occasion in fisheries in which there has been a unanimous decision of the Council. On review, in October 1984, the Council decided that the introduction should be yet again postponed, this time by a further two years, until 1 January 1987. Even this was not the end of the story. When the Commission presented its proposed regulation to replace Regulation No. 171/83 by what was to become Council Regulation No. 3094/86, the Council took the opportunity to delay its implementation yet once again, this time until 1 January 1989 with an intermediate period of two years starting 1 January 1987 during which the minimum mesh size was 85 mm. So a minimum mesh size of 80 mm eventually became law over two years later than originally agreed by the Council and 27 years after it was recommended by the Overfishing Convention. The minimum mesh size of 90 mm entered into force over six years later than originally agreed by the Council. Perhaps the greatest irony is that under an English law of 1714, the legal minimum mesh size was 90 mm, with minimum landing sizes which were for most species as large or even larger than those provided in Community legislation today.

Given the serious state of the stocks of cod and haddock in the North Sea in the mid-1980s, in 1991 the Commission proposed a minimum mesh size of 120 mm, as described in Chapter 5, the Council adopting a minimum mesh size of 100 mm which entered into force on 1 June 1992. This sequence of events is summarized in Table 6.1.

The scientific advice

As early as 1981 ACFM had recommended that the minimum mesh size in the North Sea should be increased to 90 mm. Much of the research on which this recommendation was based had been done prior to the development of the CFP. But because minimum mesh sizes was not a subject of on-going research after 1980, ACFM did not repeat its recommendation in its subsequent reports, which the Council exploited as described. When, at the request

Table 6.1 Changes in the 'standard' minimum mesh size for the North Sea.

Mesh size (mm)	Council Regulation (EEC) Nos					
	2527/80	171/83	2931/83	167/84	3094/86	345/92
70S/75D	from 01.10.80 to 30.11.80					
80	from 01.12.80 to 30.09.82	from 27.10.83 to 31.12.83	until 31.12.84	until 31.12.86	until 31.12.86	
85					from 01.01.87 to 31.12.88	
90	from 01.10.82	from 01.01.84	from 01.01.85	from 01.01.87	from 01.01.89	
100						from 01.06.92

S= single twine; D= double twine.

of the Commission, ACFM reassessed in 1984 the effects of increasing the minimum mesh size to 90 mm, it failed to made a specific recommendation. In marked contrast was ACFM's recommendations in the same report that the levels of fishing mortality on the stocks of North Sea cod, haddock and whiting should be reduced toward F_{max}, which were underlined. This must have undoubtedly given the impression that reducing fishing mortality was much more important than increasing the minimum mesh size.

The high level of discards – fish which are caught but not landed because for one reason or another they are not marketable – should also have led ACFM to recommend an increase in the minimum mesh size. From as early as 1960 it was known that about 30% by weight and 50% by number for haddock and 50% by weight and 60% by number for whiting of the fish caught in the fishery for human consumption were being discarded. ACFM strongly emphasized these high levels in its report for 1983 in which it gave the data for 1973–82, commenting that 'In view of the serious loss in yield associated with the numbers of fish being discarded, ACFM reiterates its 1981 advice for the need to increase the minimum mesh size in the North Sea as soon as possible'. But it never published the data again nor did it ever present in such clear, unequivocal language the seriousness of the situation. Contrast, for example, the language in its report for 1983 with that in its report for 1988 in which, commenting upon North Sea haddock, it stated that 'The level of discarding remains very high; reducing discards would be the most effective conservation measure, since industrial by-catch has now declined to a low level. This could be achieved by further increases in minimum mesh size and by the modification of fishing gear to reduce the retention of small fish'. The 1983 report contrasts even more with that in the report for 1989 which only briefly mentioned the problem of discards. The main emphasis in the report for 1989 was on the low level of the spawning stock biomass and the consequences for future recruitment. Even in 1991 when ACFM stated that drastic action should be taken to halt the decline in the stocks of cod and haddock, it continued to discount the importance of discards simply commenting that '. . . quantities of fish in excess of TACs were often caught and a portion of these have been discarded'. This ignored both the existence of the problem long before TACs had been used as a means of fisheries management and its gravity. Instead, ACFM recommended that fishing effort on these stocks should be reduced by 30%, a recommendation which took everyone's attention because of its implications. The problem and the importance of discards continued to be ignored.

ACFM was well aware of the situation and in its assessments of these stocks used the information on the quantities of discards and their age composition. The predictions of catch possibilities also clearly showed the estimated quantities of discards. The reason for which this issue was not highlighted can probably be explained by the manner in which the data were collected. Information on discards can be collected only by sending observers on commercial vessels, which is expensive. At that time it was being carried out only by the Marine Laboratory, Aberdeen, monitoring the Scottish fleet. There is little doubt that high rates of discarding occur in other fleets and other fisheries but, because this information appeared to lay the blame on one fleet only, the Scottish Fishermen's Federation (SFF) threatened not to co-operate with the scientists if this information was used as a basis for management decisions.

Another consideration should have caused ACFM to alert the managers to the need for an increase in the minimum mesh size and, in fact, to raise the whole question of the basis on which minimum mesh sizes were set. This was the fact that the size of haddock and whiting being caught were far too small in relation to the legal minimum mesh size. This relationship is determined by 'the selection factor', a term describing the ratio of the length at which 50% of the number of fish of any species entering the net are retained and the mesh size of the codend. Most of the work on determining selection factors was done prior to 1970. The selection factor for haddock determined from this research was 3.4 and for whiting 3.8. This means that, for a minimum mesh size of 80 mm, the length at which 50% of the haddock are retained should be 27.2 cm and for whiting 30.4 cm. The age composition data showed that much smaller fish were being caught. It is possible that ACFM simply assumed that fishermen were using illegally sized nets but there may have been other explanations. First, the assessments on which TACs are based use age, not length composition data, although compiling the latter is an essential step in constructing the former. But only the age composition data are published, removing the impact which a table of length composition makes; everyone can visualize what a 10 cm haddock looks like but not a one-year-old haddock. Second, the assessments do not require explicit account to be taken of selection factors, as did the earlier simpler models which these assessments had replaced, except when the consequences of a change in mesh size is being

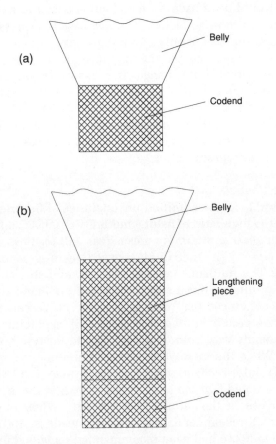

Fig. 6.1 (a) Codend attached directly to the belly of the trawl; (b) Codend attached to a lengthening piece.

considered. There was probably a third factor, namely that much of the assessment work is routine, done on computers, which leads to inconsistencies in data being overlooked.

Whatever the reasons, it is now known that developments in gear design had rendered the selection factors determined prior to 1970 obsolete. The research on which selection factors were based was done on side trawlers where the trawl, as the name implies, is hauled over the side and the codend is attached directly to the main body of the codend (Fig. 6.1 (a)). Side trawlers have all been replaced by stern trawlers. In many of these the codend is still hauled over the side but to permit this a tube of netting, called a 'lengthening piece', is inserted between the main body of the trawl and the codend (Fig. 6.1 (b)).

It is now known that the length and construction of the lengthening piece has a marked effect on selection factors. When the problem was eventually realized and further research undertaken, the results showed that the effect was to reduce the selection factor for haddock to 2.3 and that for whiting to 3.1; that is, the length at which 50% of haddock escape from a codend with an 80 mm mesh attached to a trawl rigged with a lengthening piece is 18.4 cm, not 27.2 cm as it would have been for a codend attached to a trawl rigged in the manner in which the original research was based. Expressed in another way, 80 mm old gear represents only 54 mm new gear.

Because ACFM under-emphasized the high level of discards and their effect on the stocks and overlooked the changes in selection factors, it was impossible for the Commission to convince the Council of the importance of increasing minimum mesh sizes. Instead, the reports of ACFM led the Council to believe the critical factor in management was maintaining minimum levels of spawning stock biomass. Even the manner in which this was presented led the Council to taking inappropriate action.

The actions of the Council

It is as incontrovertible as the admitted uncertainties of fisheries science permit to state that at the high rates of fishing mortality of most Community stocks of demersal fish, increasing minimum mesh sizes will result in bigger average catches in the long term. Why then did it take so long to agree to increase minimum mesh sizes? The fundamental problem was, and still is, that any proposed increase in minimum mesh size results in a predicted short-term loss. Every fisherman is convinced that these will occur but scarcely any fisherman is convinced that the predicted long-term gains will be obtained.

A series of arguments were therefore deployed to delay increases in minimum mesh sizes. When the increase in the minimum mesh size came up for review prior to 31 July 1983, as provided by Article 2 of Regulation No. 171/83, the Council noted that the latest report of ACFM made no recommendation for an increase in the minimum mesh size. When the Commission responded that such a recommendation had been made in 1981 and the reasons for which the increase had been recommended remained unchanged, the Council argued that the biological situation might have changed and that it would not act in the absence of a new assessment. This inevitably postponed taking any decision by a minimum of one year, the time it took for the request to be submitted to ICES and for a response to be obtained. It is a weakness of the Community's own scientific advisory body, the STCF, that it does not have the data banks nor the computer facilities to make such assessments. Thus, in 1984 the proposal to increase the minimum mesh size in the North Sea to 90 mm was referred back to ACFM, as previously described.

Once the advice was available, ministers concentrated upon the short term, advancing socio-economic arguments for not increasing the minimum mesh size. For example, the assessment made by ACFM in 1984 estimated that there would be immediate losses of 30% of whiting and 15% of haddock, losses which it was argued that the industry could not economically sustain. Unfortunately, ministers and the industry always equate a 10% short-term loss in weight as a 10% loss in value even though it is a readily observable fact that price elasticity of demand often results in higher prices per unit weight as supplies fall, as shown in Table 6.2. Fishermen do nothing to disabuse ministers; to do so would spoil their case.

Another economic argument employed was the cost of replacing codends. The regulations always provide for a period before the new minimum mesh

Table 6.2 Comparison of UK total landings, total value at first-hand sale and average prices of selected species to show price elasticity of demand.

Species	Weight (tonnes)		% difference	Value (£ 000s)		% difference	£/tonne		% difference
	1990	1991		1990	1991		1990	1991	
Cod	59 923	55 735	− 7	76 136	75 969	0	1 271	1 363	+ 7
Haddock	48 941	45 930	− 6	59 173	53 289	− 10	1 209	1 160	− 4
Monkfish	12 470	11 309	− 9	26 814	27 536	+ 3	2 150	2 435	+ 13
Sole	3 111	3 203	+ 3	14 700	13 790	− 6	4 725	4 305	− 9
Whiting	37 615	40 959	+ 9	25 965	22 902	− 12	690	559	− 19

Source: MAFF statistics.

sizes come into force to allow for codends to be replaced through normal wear and tear. But ministers always act as if fishermen have to use nets of the existing minimum mesh size until the date the higher one enters into force. When it is suggested that, in order to minimize the problems of short-term losses, increments should be made in steps of 5 mm, which means successive increases if realistically large increases are to be achieved, ministers argue that the industry should not have to be faced with continuous changes in the regulations and the need to buy new codends.

The Commission attempted to use the discard data, one of the most telling pieces of evidence, that an increase in the minimum mesh size was urgently needed. In the absence of specific data or recommendations in the reports of ACFM, the Commission had to refer to the reports of the ICES Roundfish Working Group. The Working Group Reports do not constitute the official advice of ICES. This placed the Commission on weak ground which was exploited by the Council. The SFF took every opportunity to deny the accuracy of the statistics, pointing out in particular the small number of samples on which they were based. The SFF emphasized instead the losses to the human consumption fisheries resulting from by-catches of haddock and whiting taken in the industrial fishery for Norway pout. All Member States, except Denmark, are opposed to industrial fishing. The Council therefore found it far easier to concentrate on this issue and ignore that of discards.

With a crisis in the fishery for North Sea cod and haddock, it is now accepted that the information on discards was valid, but too late.

When the Council eventually adopted in 1986 the proposal of the Commission to increase the minimum mesh size in the North Sea to 90 mm, it phased the increase over two years with an intermediate mesh size of 85 mm during 1987–88. This was partly to minimize the theoretical short-term losses on each increase, theoretical because most of the losses were of fish which would have been discarded, but also because the Scottish industry argued that it would be impossible to catch whiting with a minimum mesh size of 90 mm. Catches of whiting did not decrease when this minimum mesh size was eventually entered into force.

The recent history of the increase of the minimum mesh size in the North Sea is more complex in that the proposal of the Commission for an increase to 120 mm, square mesh, was not based upon any scientific recommendation. In 1991 ACFM made an assessment of the effects of implementing a minimum mesh size of 130 mm diamond mesh, which it equated with 120 mm square mesh. The results from this assessment predicted that there would be short-term losses in the human consumption fisheries of 20–31% for cod, 9–47% for haddock and 60–80% for whiting with increases in spawning stock biomasses after two years of 10–21%, 17–32% and 26–32%, respectively. It seems possible that the Commission, noting the serious concerns expressed by ACFM about the decline in the spawning stock biomasses of North Sea cod and haddock which it stated had fallen below what it considered the 'lowest desirable' levels, based its proposal on this assessment with the objective of rebuilding the sizes of the spawning stock biomasses. As described in Chapter 5, the adverse comments of the STCF did not assist the acceptance of the proposal but, despite the opposition of the fishing industry, in this case the Council did eventually adopt a minimum mesh size of 100 mm, an increase of 10 mm only three and a half years after the increase to 90 mm.

In the context of the history of increases in minimum mesh size this represented a massive increase taken in an incredibly short time. In this case ministers were more concerned about the long-term consequences of taking no action than of taking action. This reflects the fact that the basis on which the decision was made – that low spawning stock size may result in reduced recruitment and consequently a collapse of a stock – is a concept which ministers and fishermen can easily understand. It seems logical, although not true, that a stock collapse must be an inevitable consequence of low spawning stock biomass, a belief fostered by the emphasis placed by ACFM on maintaining spawning stock biomasses above critical levels. There is no stock in Community waters, however, for which a direct relationship between spawning stock biomass and recruitment has been demonstrated. The irony of this situation is that the Council ignored for years the advice of ACFM to reduce the rates of fishing mortality on these stocks. These rates still remain very high, a fact which appears to be unrecognized, perhaps because they are presented in the reports of ACFM as exponentials, not as percentage removals from the stock. Reducing these rates would have removed all the concerns about the size of spawning stock biomasses.

It is not only the scientific advice which affects what decisions are taken but also the balance of interests within the Council. Six Member States have major interests in the North Sea fisheries for demersal species. Of these, Denmark supports increases in minimum mesh sizes because its most important species are cod and saithe for which large minimum mesh sizes are appropriate. Belgium and the Netherlands are primarily interested in fishing for plaice and sole. A minimum mesh size of 120 mm or even larger would not reduce catches of marketable plaice while for sole there is a derogation permitting a minimum mesh size smaller than the standard to be used so they also support increases. Germany also falls into this category although, as described later, its position is ambivalent. The UK also has a divided position, having to balance the interests of its fisheries for haddock and, in particular, whiting against those for cod. It is essentially France which is opposed to increases in minimum mesh sizes because of its market for small whiting. This results in the situation in which, in order to obtain a qualified majority, it is often necessary to make concessions to one or other Member State, particularly those having ten votes, in order to get measures adopted.

The English Channel

The North Sea was not an isolated case. In its proposal of 1976, the Commission included the eastern English Channel with the North Sea, for which it proposed a common minimum mesh size of 90 mm. In 1981 ACFM recommended a minimum mesh size of 80 mm for all the English Channel. Despite this recommendation the Council adopted in Regulation No. 2527/80 minimum mesh sizes of 70 mm for codends braided with single twine and 75 mm for those braided with double twine. These mesh sizes were for the period 1 October to 30 November 1980. The regulation provided for the minimum mesh size to be 75 mm, irrespective of the number of twines used to construct the netting of the codend, as from 1 December 1980. The latter minimum mesh size was retained in Regulation No. 171/83, subject to a review by the Council before 31 July 1983 as to whether it should be increased to 80 mm as from 1 January 1984. The Commission proposed that this increase should take place. The Council rejected the proposal. In 1985 the Commission again proposed an increase to 80 mm. However, as part of the compromise which permitted Regulation No. 3094/86 to be adopted, the minimum mesh size remained at 75 mm. Yet once again there was a review clause which provided that '. . . in the light of scientific opinion based on further research . . .' a decision should be made 'before 1 July 1987 whether the minimum mesh size . . . should be increased from 75 mm to 80 mm from 1 January 1989'. On review, the mesh size was increased. In this case it took five years to achieve an increase of 5 mm in the minimum mesh size to 80 mm which remains the minimum mesh size.

The main reason for which the increase in minimum mesh size was so long delayed was that only two Member States, France and the UK, have a major interest in the fisheries in the English Channel. Other Member States were not prepared to support higher minimum mesh sizes to which one of the interested Member States, usually France, was opposed.

The Irish Sea

Regulation No. 2527/80 provided for minimum mesh sizes of 70 mm single twine/75 mm double twine for the Irish Sea. These were retained in Regulation No. 171/83 despite the fact that ICES had published a report showing incontrovertible scientific evidence that there was no justification for maintaining different minimum mesh sizes for single and double twine, termed a 'differential'. This was the only differential retained in Regulation No. 171/83.

In 1983, as part of a move to rationalize minimum mesh sizes, the Commission made a proposal to increase that for the Irish Sea to 80 mm, based on the similarity of the species composition of the Irish Sea and the North Sea, and to delete the differential. In the absence of specific scientific advice, the Council requested that this be obtained. ACFM examined the issue in 1984 but concluded that it '. . . considers it premature to make any recommendations for the further adjustment of mesh sizes . . .' and referred the matter back to the Irish Sea Working Group. In its report for 1985 ACFM simply reported the results of the mesh assessment without making any recommendation. In consequence, the Council refused to adopt the proposal.

In making its proposal for what was to become Regulation No. 3094/86, the Commission again proposed increasing the minimum mesh size to 80 mm and eliminating the differential. The Council agreed to drop the differential but, in the light of the assessment made in 1984 that the losses in the first year resulting from an increase in the minimum mesh size from 70 mm to 80 mm would be 16% for sole and 30% for whiting, fixed the standard minimum mesh size at 70 mm. This remained the minimum mesh size until the adoption of Regulation No. 345/92 modifying Regulation No. 3094/86 for the eleventh time by which the minimum mesh size was increased to 80 mm as from 1 June 1992.

In this case the Commission was hindered by an absence of specific recommendations from either ACFM or STCF and, in consequence, it took nine years to obtain an increase in the minimum mesh size of 5 mm, with a period of six years in which it was effectively reduced by 5 mm.

Fishing for Nephrops in the Bay of Biscay

Regulation No. 2527/80 provided that the minimum mesh size when fishing for *Nephrops* in EEC Region 3, which is mainly the Bay of Biscay, was 45 mm until 30 November 1980 and 50 mm thereafter. Regulation No. 171/83 fixed the minimum mesh size at 50 mm with a review clause identical to that already described for the Irish Sea. Each time the Commission made a proposal to increase the minimum mesh size to 55 mm when it came up for review, the Council maintained the minimum mesh size at 50 mm and provided a new date by which it should be reviewed. This series of events is summarized in Table 6.3.

In this case, the main objective of increasing the minimum mesh size was to reduce the massive catches of small hake which are taken as a by-catch in

Table 6.3 Summary of regulations concerning changes in the standard minimum mesh size when fishing for *Nephrops* in EEC Region 3 (Bay of Biscay).

Regulation No.	Decide by	Implement with effect from:	From (mm)	To (mm)
171/83	31.07.83	01.01.84	50	60
2931/83	31.12.84	01.07.86	50	60
3625/84	31.03.86	01.07.86	50	60
3094/86	31.12.87	01.01.89	50	55

the fishery for *Nephrops*. A very high proportion of these is discarded. For example, the data for 1989, which are typical of other years, show that an estimated 33 million hake were caught in this fishery. More than 18 million (55%) of these were discarded. Of those landed, seven million, representing 46% of the landings, were undersized. Despite this information and the fact that from 1987 onwards ACFM was reporting the results of assessments which showed the long-term benefits in both weight and value of increasing the minimum mesh size in this fishery it was impossible to achieve an increase. (This was the only assessment made by ACFM to include economic data.) A major problem in this fishery is the widespread use of gear having undersized meshes. ACFM had recommended that legal minimum mesh sizes should be enforced before they were increased, providing the Council with yet another excuse for not increasing minimum mesh sizes.

Eventually, in December 1989, the Council adopted a regulation increasing the minimum mesh size to 55 mm. However, the increase formed part of a 'package' of three measures, the only means whereby it was possible to get the increase adopted. The second part of the 'package' was a modification of the maximum percentage of protected species which could be caught from 60%, of which no more than 30% could be hake, to 60%, irrespective of species, a modification long sought by French fishermen. The third element made provision for the use of selective trawls, defined as those having a codend of which the minimum mesh size of the upper panel of netting was 65 mm and that of the lower panel 50 mm.

Only France has a interest in this fishery for *Nephrops* so it was easy for it to block the adoption of higher minimum mesh sizes, despite the fact that the catches of small hake reduce the catches of other Member States. However, these catches are of large hake on grounds outside the Bay of Biscay so the adverse consequences are not apparent and do not cause the same adverse response as, for example, fishing of immature herring by Denmark.

Fishing for sole in the North Sea

Regulation No. 2527/80 permitted vessels whose engine power did not exceed 300 bhp to use nets having a minimum mesh size of 70 mm single twine/ 75 mm double twine when fishing for sole in the North Sea, a provision which was retained in Regulation No. 171/83; neither regulation defined what was meant by 'fishing for sole'. The Commission made repeated proposals to delete this derogation, for two reasons. First, vessels of this engine power operate mainly in the continental coastal zone, which is the main nursery area for sole. Second, nets of this minimum mesh size retain a high percentage of sole which are less than the legal minimum landing size of 24 cm. By 1 January 1989 it had succeeded in getting the Council to delete the lower minimum mesh size of 70 mm but it was not until 1 June 1992 that the derogation was eliminated.

In this case it was Germany which opposed the proposals. Germany argued that its fisheries had been seriously damaged by the increase of the fleet of 'Euro-cutters' (see Chapter 5) and that the Commission should address this question first before taking measures which would further adversely affect its vessels. Most of these were old and low-powered and were dependent upon fishing in the coastal zone and, in particular, on catches of sole. The Commission had to modify its proposal appropriately in order to get it adopted.

Seasonal bans on fishing

Seasonal bans on fishing have the objective of protecting juvenile fish by prohibiting fishing during the periods when they are abundant in the areas where fishing is banned. The most effective measure is totally to ban all fishing. Less effective, because it makes enforcement much more difficult, is to ban all gear capable of catching the species to be protected. This section describes the history of three bans, including restrictions on fishing having the same objective, which have been largely nullified because the Council insisted on making provisions which created loopholes, rendering the provisions ineffective.

Western mackerel

The history of the regulations
As early as 1980 ACFM recommended that fishing with pelagic trawls and purse seines should be prohibited in the western end of the English Channel and the Bristol Channel (Fig. 6.2 (a)), except from mid-December to mid-February, in order to protect juveniles of the western stock of mackerel, abundant in this area. In 1982 ACFM strengthened its recommendation, enlarging the area (Fig. 6.2 (b)), making the ban applicable throughout the year and excluding all fishing gears except hand lines and gillnets. Until the settlement of the CFP there was no possibility of getting such a potentially controversial recommendation adopted by the Council, for which reason the Commission made no proposal. However, the Commission took the first opportunity after the adoption of the CFP to make a proposal. Building upon the lessons learnt from trying to enforce the provisions of Regulation No. 171/83, a total ban on fishing within the area, except for hand lines and gillnets, was proposed. This proved unacceptable to the Council for the reason that several Member States wanted to fish for horse mackerel, pilchards and sprats in this area. In order to get a measure adopted, the Commission had to modify its proposal but succeeded in getting provisions adopted which meant that a vessel having more than 15% of mackerel on board was deemed guilty of fishing for mackerel unless the quantities on board had been reported prior to its entry into the restricted area. Provision also had to be made to exempt vessels fishing with nets having a minimum mesh size equal to or larger than 70 mm, the minimum mesh size for *Nephrops*. The Council also fixed the eastern boundary at 3°W instead of 2°W as recommended by ACFM and proposed by the Commission and also provided that the legislation would expire on 1 January 1987 unless it decided otherwise before 31 July 1986. This legislation was adopted by Regulation No. 2931/83 of 4 October 1983, the first amendment to Regulation No. 171/83.

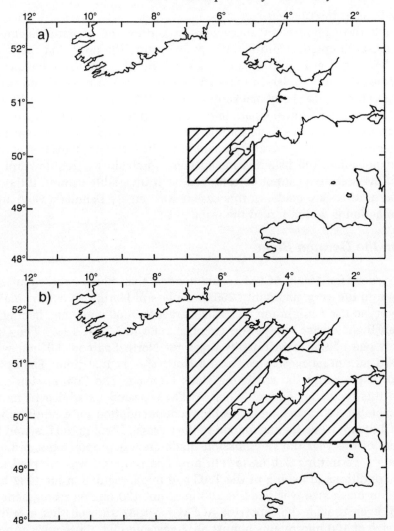

Fig. 6.2 The 'mackerel box', the area in which fishing for mackerel is restricted (see text): (a) as originally recommended by ACFM in 1980; (b) as subsequently recommended in 1982.

In 1988 the Commission proposed once again that the eastern boundary of the prohibited area should be fixed at 2°W. This time the Council adopted the proposal. When the legislation became subject to renewal in 1986 the Commission proposed its continuation. The Council adopted the proposal but again insisted on the insertion of a review clause providing for the legislation to expire on 1 January 1992 unless it decided otherwise by 30 November 1991. In 1991 the Commission proposed that the renewal provisions should lapse, which the Council adopted in the tenth modification of Regulation No. 3094/86.

The actions of the Council

Although the Commission succeeded in getting the essential elements of its proposal adopted, it did so only by accepting the provision that a vessel could have on board up to 15% of mackerel. But this provision has permitted vessels legitimately to fish in the area for sprat, horse mackerel and pilchard, as long as they discard any mackerel over the 15% limit, largely nullifying the measure. This is a typical example of how proposals to ban fishing in certain areas in certain seasons have foundered upon the rock of self-interest of the Member States. It is a general argument made by Member States that such bans should not affect the fisheries for species which ·do not require protection, even if the objective cannot be achieved, as it invariably cannot, if the necessary derogations are made. In this case it was largely Denmark, Germany and the Netherlands which forced the issue.

Cod in the German Bight

In its report for 1986, ACFM recommended that 'some protection' should be given to the very abundant 1985 year-class of North Sea cod by making it obligatory to use a minimum mesh size larger than 80 mm when fishing in the German Bight during the period October 1986 to March 1987. The German Bight is one of the main nursery areas for North Sea cod; 80 mm was the standard minimum mesh size for the North Sea at that time. The objective was to help rebuild the spawning stock biomass. The Commission did not act on this recommendation. When ACFM repeated its advice in its report for November 1987, making a specific recommendation for a minimum mesh size of 120 mm when fishing for cod in the German Bight in the first and fourth quarters of the year the Commission made such a proposal but provided a derogation permitting fishing for shrimp. The proposal was adopted as part of the conditions of fishing in the TAC and quota regulation for 1987 but the minimum mesh size was fixed at 100 mm, not 120 mm as recommended by ACFM, even though the report of ACFM for 1987 showed that a minimum mesh size of 100 mm retains almost all 2-year-old cod, those which the measure was supposed to protect (Table 6.4).

In its report for 1987 ACFM recommended the continuation of this measure, again advising a minimum mesh size of 120 mm. Again the Commission proposed the use of 120 mm and again, for exactly the same reasons the Council adopted the box with a minimum mesh size of 100 mm plus a derogation permitting the use of nets having a minimum mesh size of 16 mm when fishing for eels. The regulation also made it compulsory to use a separator trawl when fishing for shrimps. A separator trawl has a horizontal sheet of netting in the codend which divides the codend in two; this sheet of netting has a minimum mesh size which is related to the smaller of the two species being caught, in this case 20 mm for shrimps. The upper panel of the codend is made of the standard mesh size, in this case 100 mm. This scenario has been re-enacted every year since, although the entry into force of a minimum mesh size of 100 mm in the North Sea as from 1 June 1992 made the measure redundant and it was not included in the TAC and quota

Table 6.4 Percentage retention of one- and two-year-old North Sea cod by mesh size.

Age (years)	Mesh size (mm)				
	80	90	100	120	140
1	46	18	6	1	0
2	100	99	97	67	21

Source: Report of ACFM for 1987.

regulation for 1993. A minimum mesh size of 120 mm in this area is still unacceptable.

The case of fishing for cod in the German Bight was complex because the initial scientific advice had the specific objective of protecting the 1985 year class of cod. By the time that the Commission made its proposal, the measure was too late to achieve this objective. Germany used this argument to oppose the proposal. However, its main objective was to protect its coastal fleet which was heavily dependent upon this fishery, the location of the base ports and the size of the vessels involved preventing them from fishing elsewhere. Germany also argued that juvenile cod occurred abundantly elsewhere and that the restriction should also apply in these areas if it was to be an effective conservation measure. This was also a political argument; German fishermen should not be discriminated against. These other areas were mainly off the northeast coast of England and this was an attempt to get the support of other Member States. The support of Germany was needed to obtain a qualified majority in favour of the TAC and quota regulation, of which this condition of fishing formed part, and, as a compromise, the minimum mesh size was reduced from 120 mm, as recommended by ACFM, to 100 mm. The derogation for shrimp had already been proposed by the Commission because the Commission realised that it would be impossible to gain acceptance of a proposal stopping this important fishery.

Again this case illustrates the power which a Member State having ten votes can wield if these are essential for the adoption of a proposal. This case also illustrates another facet of how the Community operates. The minimum mesh size of 100 mm had no conservation benefit yet the Commission did not withdraw its original proposal and the Council adopted the compromise proposal. The reasons were nothing to do with conservation. The former course would have been regarded as a political defeat for the Commission while the latter had the political advantage of demonstrating that the Council was apparently taking conservation measures.

North Sea herring and sprat

One of the major nursery areas for North Sea herring is off the west coast of mainland Denmark. In order to prevent fishing for juvenile herring in this area, the Council adopted as part the conditions of fishing in the TAC and quota regulation for 1983 bans on fishing for herring and sprat in this area during the period 1 July to 31 October; the ban on fishing for sprat had the objective of preventing fishing for juvenile herring under the pretence of fishing for sprat. This provision has been adopted every year since and was transferred to Regulation No. 3094/86 by its eleventh modification, making the bans permanent.

These bans illustrate how the interests of a small Member State, having few votes, can be overridden, the reverse of the situation when the votes of a large Member State are required. These bans had as serious an effect on the Danish vessels fishing for small clupeoids as the proposal to use a minimum mesh size of 120 mm in the German Bight potentially had on German vessels, for the same reasons. However, in this case the fishery was considered largely responsible for the collapse of the stock of North Sea herring and its failure to rebuild; all Member States, including Denmark had an interest in re-establishing this stock because of the importance of the fishery based upon it. Also, the catches of juvenile herring were used for reduction to meal and oil, anathema to all Member States except Denmark. Realizing that the cards were stacked against it and that it had no hope of preventing the ban, Denmark did not object. However, by taking this action it achieved a political objective. Denmark's negotiating stance towards the end of the debate on the conservation policy had been particularly obdurate and had delayed its settlement beyond the end of 1982. As a result, Denmark's position in the Council was politically very sensitive. By not opposing these bans Denmark improved its political standing, which was to its advantage in negotiations on other issues. The enforcement of the bans has caused Denmark considerable political problems with the sector involved. On at least one occasion riot police have had to be used in order to quell fishermen violently demonstrating against the bans. Also, the families of the inspectors were intimidated causing Denmark to reorganize its inspectorate. Inspectors are no longer based in the towns in which they have to carry out their duties.

Total allowable catches

North Sea plaice

There are a large number of stocks for which the Council has each year consistently fixed TACs higher than those recommended by ACFM, of which North Sea plaice is but one example (Table 6.5). From 1984 to 1986 ACFM recommended that the TACs should correspond to a reduction in the rate of fishing mortality towards F_{max}. From 1987 to 1990, ACFM recommended that rates of fishing mortality should be fixed at levels which would maintain the spawning stock biomass above 300 000 t, which it considered the minimum desirable level. The TACs corresponding to all these recommendations were ignored. As a result, the rate of fishing mortality on this stock has increased year after year (Table 6.5) causing ACFM to comment in its report for 1991 that 'Although the SSB (spawning stock biomass) is well above the

Table 6.5 North Sea plaice: comparisons of (1) the TACs recommended by ACFM and those fixed by the Council and (2) the equivalent rates of fishing mortality corresponding to the TACs recommended and fixed, and also the actual rates of fishing which occurred. (TACs in thousands of tonnes and rates of fishing mortality in percentages.)

Year	TAC		Rate of fishing		
	Recommended	Fixed	Recommended	Fixed	Actual
1984	150	182	21	27	34
1985	130	200	25	na	34
1986	<160	180	34	na	40
1987	120	150	34	42	38
1988	150	175	30	36	43
1989	<175	185	38	45	41
1990	171	180	36	38	43
1991	169	175	42	53	36
1992	135*	175	36	46	(37)
1993	170	175	37	38	na

< *maximum TAC recommended*: in most years ACFM recommended that the TAC should be fixed lower than the TACs shown.
* For 1992 ACFM recommended TACs corresponding to a reduction in fishing effort of 20%; the TACs shown correspond to a reduction in the rate of fishing mortality by 20%.
The actual rate of fishing mortality for 1992 is provisional.
na: data not available.
Sources: Reports of ACFM and Council regulations fixing TACs.

target minimum level of 300 000 t, fishing mortality is at a historic high level. There is no benefit to be gained in terms of long-term yield by increasing fishing mortality above the present level.' Despite this, the Council adopted for 1992 a TAC which represented a 10% increase in the rate of fishing mortality.

As described earlier, once ACFM had set a minimum spawning stock biomass as the biological management objective it was possible for the Council to argue that the TAC could be fixed higher than that recommended by ACFM if to do so would not result in a spawning stock biomass less than the target level recommended by ACFM. The Netherlands, the Member State with the greatest interest in this stock, consistently advanced this argument.

North Sea roundfish

The history of fixing TACs for North Sea roundfish is similar to that for North Sea plaice but has the special feature of decisions being taken which nullified the objectives which the Council thought it had set out to achieve.

Cod
Up until and including 1986, ACFM recommended that TACs should be fixed which corresponded to reductions in the rate of fishing mortality towards F_{max},

Table 6.6 North Sea cod: comparisons of (1) the TACs recommended by ACFM and those fixed by the Council and (2) the equivalent rates of fishing mortality corresponding to the TACs recommended and fixed, and also the actual rates of fishing which occurred. (TACs in thousands of tonnes and rates of fishing mortality in percentages.)

Year	TAC		Rate of fishing		
	Recommended	Fixed	Recommended	Fixed	Actual
1984	<182	215	<56	58	57
1985	<259	250	<60	58	56
1986	<130	170	<47	63	59
1987	<125	175	<50	63	59
1988	≤148	160	≤47	52	59
1989	<124	124	<50	50	62
1990	113	105	47	44	55
1991	92*	100	46	50	61
1992	82*	100.7	42	53	(61)
1993	92*	100.7	48	52	na

< *maximum TAC recommended*: in most years ACFM recommended that the TAC should be fixed lower than the TACs shown; for example, for 1989, ACFM expressed a preference for a TAC of 100 000 t.
* For 1991, 1992 and 1993 ACFM recommended TACs corresponding to a reduction in fishing effort of 30%; the TACs shown correspond to a reduction in the rate of fishing mortality by 30%.
The actual rate of fishing mortality for 1992 is provisional.

Sources: Reports of ACFM and Council regulations fixing TACs.

which was equivalent to a rate of fishing of 17% a year compared with actual rates of about 65%. As for North Sea plaice, the advice was ignored. In 1984, 1986 and 1987, the Council adopted TACs which corresponded to increases in the rate of fishing (Table 6.6). From 1988 onwards the Council did adopt TACs which corresponded with reductions in the rate of fishing. But even in 1991 and 1992, when ACFM was stressing the urgent need to reduce fishing effort by 30%, the TACs adopted corresponded with much smaller reductions than those advised. As a consequence of not implementing the recommendations of ACFM the rate of fishing mortality in 1991 was 61% a year, higher than that at the start of the CFP. The TACs have achieved nothing.

Haddock

The situation for haddock was similar to that for cod. From 1984 to 1987, ACFM recommended that TACs should be fixed which corresponded to a reduction in the very high rate of fishing mortality, of the order of 65% a year. As usual, the Council fixed TACs which corresponded with maintaining the rate of fishing mortality at its existing rate, the *status quo* option, except notably in 1987 when it corresponded with a 25% increase (Table 6.7). Since 1988 the TACs adopted have respected the scientific advice but despite this the rate of fishing mortality remains close to that in 1984 – 60% a year compared with the target of 50% to which the TAC for 1992 corresponds.

Table 6.7 North Sea haddock: comparisons of (1) the TACs recommended by ACFM and those fixed by the Council and (2) the equivalent rates of fishing mortality corresponding to the TACs recommended and fixed, and also the actual rates of fishing which occurred. (TACs in thousands of tonnes and rates of fishing mortality in percentages.)

Year	TAC		Rate of fishing		
	Recommended	Fixed	Recommended	Fixed	Actual
1984	<162	170	<63	67	63
1985	<209	207	<56	56	61
1986	<239	230	<57	57	66
1987	<120	140	<66	80	64
1988	<185	185	<57	57	65
1989	<68	68	57	57	60
1990	50	50	55	55	55
1991	48*	50	49	49	71
1992	61*	60	48	48	(69)
1993	158*	133	56	50	na

< *maximum TAC recommended*: in most years ACFM recommended that the TAC should be fixed lower than the TACs shown.
* For 1991, 1992 and 1993 ACFM recommended TACs corresponding to a reduction in fishing effort of 30%; the TACs shown correspond to a reduction in the rate of fishing mortality by 30%.
The actual rate of fishing mortality for 1992 is provisional.
Sources: Reports of ACFM and Council regulations fixing TACs.

Whiting

For 1984 and 1985 the Council adopted TACs which were so high that they fell outside the highest limits of the scientific advice. They would have required rates of fishing mortality greater than 90% a year (Table 6.8). This was done for political reasons, to provide apparently good catch possibilities in order to ensure that the TAC and quota regulation were adopted without difficulties, for the reasons explained in Chapter 4. In 1986 and 1987 the TACs corresponded with the *status quo* option compared with the recommendations of ACFM that rates of fishing should be reduced towards F_{max}. From 1988 to 1991 the TACs corresponded to, or were even less than the scientific advice but in 1992 the TAC again corresponded with the *status quo* option instead of the 30% reduction advised. As for the other two stocks, the rate of fishing has fallen only slightly, now being 55% a year compared with 60% a year at the start of the CFP.

'Tie-ups'

The recommended reductions in the TACs for cod from 1987 onwards and for haddock from 1988 onwards were based upon the urgent need seen by ACFM to rebuild the spawning stock biomasses of these two stocks which were falling to below what ACFM considered critical levels. In response to the advice of ACFM, the Council adopted an additional measure to reinforce the

Table 6.8 North Sea whiting: comparisons of (1) the TACs recommended by ACFM and those fixed by the Council and (2) the equivalent rates of fishing mortality corresponding to the TACs recommended and fixed, and also the actual rates of fishing which occurred. (TACs in thousands of tonnes and rates of fishing mortality in percentages.)

Year	TAC		Rate of fishing		
	Recommended	Fixed	Recommended	Fixed	Actual
1984	<86	149	<62	89	58
1985	<118	160	<60	58	(1)
1986	<135	135	<55	55	59
1987	<127	135	<52	53	67
1988	<134	120	<57	49	56
1989	<115	115	<42	42	56
1990	130	125	48	45	54
1991	134*	141	37	42	62
1992	120*	135	42	54	(62)
1993	96*	120	49	72	na

< *maximum TAC recommended*: in most years ACFM recommended that the TAC should be fixed lower than the TACs shown.
For 1991, 1992 and 1993 ACFM recommended TACs corresponding to a reduction in fishing effort of 30%; the TACs shown correspond to a reduction in the rate of fishing mortality by 30%.
The actual rate of fishing mortality for 1992 is provisional.
(1) The TAC corresponded with a rate of fishing mortality exceeding 95%.
Sources: Reports of ACFM and Council regulations fixing TACs.

TACs for 1991 and 1992 and reduce fishing effort by 30%. As part of the conditions of fishing, provisions were adopted under which vessels whose landings traditionally consisted of cod and haddock had to spend eight consecutive days in port each calendar month from February to December in 1991 and 1992 or, alternatively for 1992, a total of 135 days in port during the same period. In each year provision was made for the Commission to exempt vessels if it considered that a Member State would be unable to catch its quotas. These provisions also applied to fishing west of Scotland, where the stocks of cod and haddock were also decreasing rapidly.

In a situation in which the fleets are so large that quotas can be exhausted before the end of the year if vessels are allowed to fish without restriction, limiting the number of days at sea is the most effective means to achieve quickly the objective of reducing rates of fishing mortality. That it is so effective can be gauged by the strong opposition of the UK fishing industry to the eight-day tie-up during 1991, which it managed to get reduced from the ten days originally proposed by the Commission, and the present opposition to the UK government's Sea Fisheries (Conservation) Bill which permits it to limit days at sea for all UK registered vessels. When the Commission proposed that there should be a tie-up period of 200 days in 1992, the UK fishing industry, which was potentially the worst affected by the proposal, bitterly opposed it. As a result the Council not only reduced the period to 135 days but made no provision for these to be consecutive, meaning that all days in port for whatever reason could count. Taking account of days in port between trips and for maintenance this effectively meant little or no reduction in fishing effort because many vessels spend as much, if not more time in port than 135 days. For example, the two vessels based in Lowestoft with the highest earnings in 1991 spent 105 and 111 days in port. It is also illustrative that the UK, at the instigation of the Commission, introduced a limit of 92 days in port in 1990, also without any provision that these should be consecutive and that this produced very little opposition, obviously for the reason that it had little or no effect on normal fishing operations. This clearly illustrates both the effectiveness of the fishing industry as a lobby and the unwillingness of the Council to withstand it.

The regulations made provisions to exempt vessels if it considered that a Member State would be unable to catch its quotas. Under these provisions, UK vessels whose owners agreed to use nets having a minimum mesh size of 110 mm did not have to restrict their numbers of days at sea at all in 1991. For 1992, those using a minimum mesh size of 110 mm had to spend only 67 days in port while those using a minimum mesh size of 120 mm were totally exempted. These exemptions totally ignored that the recommendations of ACFM were for a reduction in fishing effort on all age groups, not simply on the youngest. In the case of cod it has already been noted that a mesh size of 120 mm is the minimum which provides any protection for fish less than two years old so, to the extent that vessels used minimum mesh sizes of 100 mm in 1991, there was no reduction in fishing effort on cod whatsoever as a result of this measure. It is also doubtful whether there was any reduction for haddock because there were no provisions in the regulations for 1991 to prevent the fishing gear being rigged in a manner which circumvented the effects of any increase in minimum mesh size. Such measures were introduced on 1 June 1992 so the exemptions may not have been too detrimental in 1992, especially as the stock of haddock consisted mainly of fish aged one- and two-years old in 1992, small enough to escape through meshes of 110 mm and 120 mm. However, it is very doubtful whether the measure had any conservation effect for cod. These decisions illustrate how little the scientific advice is understood.

That the TAC for whiting for 1992 was fixed at a level which corresponded with no reduction in the rate of fishing mortality and that for cod in a reduction of only 5% is also likely to have undermined the attempt to reduce fishing on cod and haddock in 1992, particularly for haddock. Under the Community system of managing TACs and quotas, fishing can continue on cod and whiting until the quotas are caught even though this may result in catches of haddock long after the quotas for this species have been exhausted. The Scottish fishing industry has always argued that it is impossible to prosecute a directed fishery for whiting without catching the other two species, in particular haddock.

For 1993 the UK argued successfully that its Sea Fisheries (Conservation) Bill which provides for limiting days at sea for all UK vessels irrespective of where they fish obviated the need for Community provisions for the North Sea and west of Scotland. With the argument over 'subsidiarity' fresh in its ears, the Commission was doubtless in no mood to argue against the UK.

English Channel cod

The Council is always prepared to bow to the demands of the fishing industry to set the highest possible TACs. Although described for convenience as English Channel cod, the TAC for this stock includes all cod caught in that area, in the Celtic Sea and to the west of Ireland, ICES sub-area VII, excluding the Irish Sea. However, the main story concerns fish caught in the English Channel so the title is appropriate. The TAC for this stock is not based upon an analytical assessment but upon an amalgam of catch predictions using simple assessment models for some components of the stocks and catch statistics for others; for this reason it is known in Community terms as a 'precautionary TAC'. The rate of fishing mortality on this stock is estimated to be very high, of the order of 70% a year, and landings therefore consist of predominantly small, young fish. Annual catches reflect almost directly the number of young fish which enter the fishery each year, the 'recruits'.

In 1987 landings started to increase dramatically and the TAC which had been fixed became limiting. As a result, fishing by some Member States, particularly the UK and France, was stopped when they exhausted their quotas. As a consequence, the Council demanded that the STCF be convened to reassess the stock. The advice of STCF permitted the TAC to be increased from 16 000 t to 19 000 t. The same situation occurred in 1988, the original TAC of 22 000 t being increased to 23 900 t. It happened once again in 1989 when the TAC was increased twice, from 23 000 t originally to 24 600 t and finally to 25 400 t.

This illustrates how pressure from the industry can result in TACs being rapidly increased. In this case the entry of two exceptionally large year-classes of cod into the fishery resulted in catches increasing rapidly and quotas becoming exhausted. The fishing industry argued that it was unacceptable that it should be prevented from catching fish which were there, or worse, being forced to throw back fish because quotas had been exceeded. The fishermen reinforced their argument by pointing out that the TAC was only precautionary and it was therefore unacceptable for them to be prevented from catching fish in the absence of firm scientific advice on which to base the TAC. The Commission argued in vain that the high catch rates were the result of two large year-classes which respecting the TACs would husband, enabling their benefits to be gained over several years. The Commission also pointed out that fishermen were not forced to go to sea to catch fish and that glutting the market with large landings would result in lower prices per unit weight and no increase in earnings. None of these arguments prevailed. The ministers involved, those of France and the UK, supported their fishermen in order to avoid the political problems involved in stopping fisheries once quotas had been exhausted. This led to the Commission, after requesting advice from STCF, to make the proposals to increase the TAC, which were adopted without any delay. The increases have almost certainly had adverse long-term consequences for the UK fishing industry by encouraging the construction of vessels less than 10 m long which were not subject to the UK system of pressure stock licensing before 1993. Once the two large year classes had been fished out, which happened very rapidly because the TACs were increased, the new vessels were left with very limited catch possibilities. It is noteworthy that since this event there have been considerably more complaints about the low level of the UK quotas in this stock compared with that of France than there were before.

Political expediency

All these case histories except two exhibit a common feature. The Commission made proposals which the Council adopted either only after a long delay or in a very modified form, often including provisions which negated the objective of the scientific recommendation on which they were based. No wonder that the view 'from Brussels' of the fishing industry is as Fig. 6.3 – compare this with Fig. 1.1! Why, given that Regulation No. 170/83 provides that conservation measures '. . . shall be formulated in the light of the available scientific advice . . .' has the Council ignored the scientific advice or at best paid lip-service to it? The overriding reason, illustrated by the factor common to all the examples describing attempts to increase minimum mesh sizes, implement effective seasonal bans on fishing and decrease TACs, is that the Council is reluctant, even opposed, to adopting regulations which result in short-term adverse economic consequences for the fishing industry. These are, of course, the consequence of every conservation measure because they are proposed only when there is a problem; adopting measures in order to avoid creating a problem is politically not even an option to be considered. When fish are abundant, ministers argue that there is no need to take action. When fish become scarce, ministers delay action in response from pressure from their fishing industries which argue that the proposed measures will only increase the economic difficulties from which they are already suffering; ministers hope that a natural recovery in the size of the stocks will resolve the problem. If this does not occur, the economic situation of the industry becomes even worse, making it even more difficult to take a decision and leading to further procrastination. Only when ministers consider that the political consequences of taking no action will be worse than those of taking action do they act. But even when difficult decisions are finally forced upon them, they are unfortunately still prepared to make provisions which nullify the intended objectives. This contrasts with the situation in which implementing the scientific advice results in short-term benefits, in which case the Council is prepared to act immediately.

Regulation No. 170/83 provides ministers with a legal basis for using 'socio-economic reasons' for not adopting proposals, postponing a decision or taking a less drastic measure than that proposed by the Commission. Its Article 1 provides that the objectives of the conservation policy are, *inter alia*, the 'balanced exploitation of (the biological resources of the sea) on a lasting basis

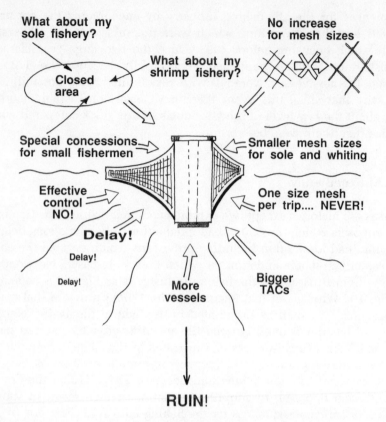

Fig. 6.3 The view from 'Brussels'; compare this with Fig. 1.1.

and in appropriate economic and social conditions'. As described, the Commission tried unsuccessfully to use economic arguments to support its proposals but was unable to support them because the scientific reports rarely include economic assessments which it could cite; this is a major deficiency of the advice available to the Commission. In consequence, its arguments were dismissed. In contrast, no such onus to substantiate their reasons for rejecting proposals is placed upon ministers who never have to specify their socio-economic reasons; the phrase can be uttered as a magic talisman which is accepted by every minister present because it provides a ready justification for any of them to vote against a proposal; no minister wants to destroy the magic formula which he may wish to use on another occasion.

But given that Regulation No. 170/83 also provides that conservation measures '. . . shall be formulated in the light of the available scientific advice . . .', are there other reasons why it has been so consistently ignored when decisions were taken? There were reasons for the Commission proposing TACs corresponding to maintaining rates of fishing mortality at existing levels. As explained in Chapter 3, this was a major factor in permitting the conservation policy to be settled. Subsequently, the events of 1983–89, described in Chapter 4, militated against the Commission proposing TACs which corresponded to their reduction. But what were the reasons for which TACs have been consistently fixed which corresponded to increases in the rates of fishing mortality, not only for the stocks described in this chapter but for many others, examples of which are given in Chapter 7?

Part of the reason lies in the absence of long-term objectives. This means that the Council takes its decision each year in the context of the existing political situation. Part lies in the fact that the scientific advice is given in the absence of objectives set by the 'managers', resulting in such advice as 'to reduce the rate of fishing mortality to F_{max}', whose implications for the long-term beneficial management of the stocks ministers may not realize and which does not provide them with specific tonnages. Also, as described, in certain cases ACFM failed to make recommendations or when they did, formulated them in a manner which did not convey the urgency of the situation. Clearly, the different type of advice of ACFM concerning the stocks of North Sea and west of Scotland cod, haddock and whiting in 1991 and 1992 and its refusal to recommend TACs convinced the Council that the situation was seriously amiss and jolted it into action. Yet the writing had been on the wall since the start of the CFP that the situation which exists today could have occurred at any time previously because the high rates of fishing on the stock have meant that two consecutive small year classes would have serious consequences. However, given the ability of the Council to delay taking decisions it is unlikely that it would have responded even if ACFM had pointed out the dangers in forceful, unequivocal language.

The increasing emphasis which ACFM has placed upon maintaining a minimum level of spawning stock biomass, not only for this stock but for others, has also been a very important factor. The majority of politicians, administrators and fishermen, and even some scientists, now consider that this is the only objective which should determine the size of TACs. The concept of maintaining a large spawning stock biomass in order to maintain high recruitment is deceptively simple and therefore easy to grasp, much easier than that of fishing mortality rates expressed as exponentials and the concept of yield per recruit. Even for those stocks for which ACFM has not recommended a minimum level, it has become increasingly common to argue that a TAC which does not result in a decline in the spawning stock biomass from one year to the next is justified, irrespective of rates of fishing mortality.

However, while the scientific community can be criticized for the manner in which it gives its advice, the core of the problem is the unwillingness of most fisheries administrators and those in the fishing industry to attempt to understand the basic biological facts underlying the advice. Without such an understanding it is impossible to evaluate the advice correctly. Decision taking will continue to be seriously flawed until this occurs.

Chapter 7

Evaluating the Conservation Policy

or 'Crisis, What Crisis?'

The criteria for evaluation

The conservation policy is considered almost universally to have been a disastrous failure. This chapter examines the extent to which this conclusion is valid, if at all, and the reasons for this. In order to evaluate the policy it is first necessary to establish the criteria by which it should be judged. In theory this should be easy, it being necessary only to compare the results achieved with the objectives of the policy. In fact it is far less simple than this because there were in reality two sets of objectives, one of which was political and the other practical. The two objectives did not coincide. Furthermore, while some of the political objectives were explicit, stated in Regulation No.170/83, others were implicit, only to be deduced from the manner in which the conservation policy was negotiated.

Evaluation has been facilitated by the adoption on 20 December 1992 of Council Regulation (EEC) No. 3760/92 establishing a Community system for fisheries and aquaculture, which replaced Regulation No. 170/83. It can be assumed that those elements of Regulation No. 170/83 which were retained unchanged in the new regulation are considered to have been politically successful by the Member States. This is particularly the case for those elements for which the Commission proposed significant changes in its proposal for the new regulation and which were not adopted by the Council.

The practical implications were even less explicit and were those which the fishing industry thought that 'conservation' should and would achieve: bigger catches, higher catch rates and greater prosperity. Many of the problems which confront the conservation policy and the main reason for which the fishing industry considers that the CFP has been a failure result from the divergence between the political and practical objectives.

Political objectives

Regulation No. 170/83

The first paragraph of Article 1 of Regulation No. 170/83 describes the objectives of the conservation policy as:

'In order to ensure the protection of fishing grounds, the conservation of the

116

biological resources of the sea and their balanced exploitation on a lasting basis and in appropriate economic and social conditions, a Community system for the conservation and management of fishery resources is hereby established.'

The phraseology is so imprecise as to be almost meaningless and is certainly of no use in providing criteria by which to evaluate the policy. What did the Council mean by 'protection of the fishing grounds'? Even more open to question is what was intended by 'balanced exploitation . . . in appropriate economic and social conditions'. The wording suggests that each Member State had its own particular objectives and that Article 1 represents a compromise which embodies all of them, even though, as will be analyzed in Chapter 12, many of them are incompatible. The compromise was necessary to obtain agreement, otherwise there might not have been a conservation policy at all. What the Member States actually saw as the political objectives can be deduced from the components of the policy and the manner in which it was negotiated and implemented.

Access

As described in Chapter 3, one of the major objectives in negotiating the conservation policy was resolving the issue of access. The agreement reached was unsatisfactory to the Member States who were mainly concerned; for France, it represented an unwarranted derogation from the provisions of unrestricted access as provided in Regulation No. 101/76 whereas for Ireland and the UK it provided far less than either wanted. However, because the agreement on access was less than satisfactory to these Member States does not mean that its implementation has been a failure. Two factors enable its success or failure to be evaluated. First, whether it has given rise to problems. Second, whether the provisions were modified at the ten-year review provided by Articles 6, 7 and 8 of Regulation No. 170/83.

Access has not given rise to political problems since 1983. In the lead-up to the ten-year review, Ireland tried to improve its position by arguing that access to its coastal zone should be restricted exclusively to vessels registered in Ireland. Also, the Shetland Fishermen's Association, having complained for many years that the 'Shetland box' had not operated satisfactorily, lobbied hard for the number of vessels licensed to fish in the 'Shetland box' to be reduced and to have it enlarged. But the Association's case was opposed by all other UK fishermen and was never part of the UK government's official position. These were attempts to use the window of opportunity offered by the review to change the conditions of access. They represented dissatisfaction by the parties concerned but access was essentially a political success from 1983 to the end of 1992. The key criterion by which this can be judged is that access has never once been the subject of debate in the Council.

This is confirmed by considering the result of the ten-year review. The Commission, in its proposal for what was to become Regulation No. 3760/92, used different phraseology to that in Regulation No. 170/83. The Commission also proposed a paragraph which could have been interpreted as restricting access, giving Member States the possibility to limit fishing in their 12-mile coastal zones to vessels operating '. . . from ports in that geographical coastal area'. Perhaps this was intended as a concession to Ireland's position. Whatever the situation, the provisions and the phraseology concerning access in both the preamble and relevant article in Regulation No. 3760/92 are almost identical to those in Regulation No. 170/83. Obviously the Council wanted to ensure that there was no possibility of the provisions on access agreed in 1983 being modified. Changing the phraseology could well have opened this possibility.

The provisions on access were virtually unquestioned for ten years, from 1983 to 1992. The Council adopted them unchanged to continue for another ten years, until 31 December 2002. Politically, access must be adjudged to have been a success.

Relative stability and Hague Preferences

The legal background

The other keystone of the conservation policy is relative stability, provided for in paragraph 1 of Article 4 of Regulation No. 170/83 which states:

> 'The volume of the catches available to the Community referred to in Article 3 shall be distributed between the Member States in a manner which assures each Member State relative stability of fishing activities for each of the stocks considered.'

This article has to be read in conjunction with three of the recitals in the preamble to the regulation which state:

> 'Whereas conservation and management of resources must contribute to a greater stability of fishing activities and must be appraised on the basis of a reference allocation reflecting the orientations given by the Council;
>
> Whereas, in other respects, that stability, given the temporary biological situation of stocks, must safeguard the particular needs of regions where local populations are especially dependent on fisheries and related industries as decided by the Council in its resolution of 3 November 1976, and in particular Annex VII thereto;
>
> Whereas, therefore, it is this sense that the notion of relative stability aimed at must be understood; . . .'

The language is far from precise but one point is clear. The 'reference allocation reflecting the orientations given by the Council' is the percentage allocation of the EC share of the TAC for each stock as fixed in 1982 for the stocks listed in Regulation No. 172/83 fixing the TACs and quotas for 1982 or as subsequently determined for other stocks. The 'orientations of the Council' were those determining how the criteria of historic catches, Hague Preferences and compensation for jurisdictional losses should be applied, as described in Chapter 3.

Relative stability

Only one permanent change has ever been made to the allocation keys agreed on 25 January 1983, Belgium obtaining an allocation in the west of Scotland and Rockall stock of haddock, as described in Chapter 4. Some of the allocation keys agreed in the first TAC and quota regulation following the accession of Spain and Portugal were also changed subsequently but this was a result of difficulties in interpreting how the provisions of the Act of Accession should be applied, not matters of principle.

Spain and Portugal have constantly queried the application of the principle of relative stability to stocks for which they consider that they should have large allocations, notably those in the waters of third countries. Between them, they brought 17 cases before the Court of Justice in the period 1990–92 concerning the allocation of quotas in the waters of Greenland, the Faroe Islands, Norway and Sweden. One of the main arguments adduced to support their cases was that the rigid maintenance of percentages fixed for certain Member States was based upon a misinterpretation of the principle of relative stability. Both Spain and Portugal are particularly incensed by the fact that prior to 1977 both caught large quantities of fish in what are now Greenlandic waters and that the Member States to which quotas are allocated in these waters fail to catch them.

In its judgement, the Court dismissed all the cases. The primary reason was that, by Article 2 of the Act of Accession, Spain and Portugal had accepted the *acquis communitaire*, which included Regulation No. 170/83 defining relative stability, and that the Act of Accession made no provision for the allocation keys to be modified.

The Court also used the argument that relative stability is based upon percentages, not upon catches and that there is no provision for the percentages to be revised as catch possibilities change, as Spain and Portugal had contended. The Court ruled that fishing activities prior to 1977 were irrelevant and, perhaps more interestingly, that the fact that the Member States failed to catch all their quotas was not a result of their under-utilization but because the agreement included quantities of fish which were simply not available. The Court did rule that the percentages could be changed by a decision of the Council, which is self-evident and provided for by Regulation No. 3760/92.

In reaching its decision, the Court drew upon its ruling made in 1987 that, even though the principle of relative stability derogates from that of equal access, it is not discriminatory between nationals because it shares the burden of insufficient resources in a non-discriminatory manner between Member States. This earlier case concerned the prosecution by the Dutch authorities of a Dutch fisherman called Romkes who had continued to fish after the Dutch quota for North Sea plaice had been exhausted. Romkes argued from Regulation No. 170/83 that the TAC, not the quota, was the conservation measure. The TAC had not been exhausted, nor was it probable that it would have been. Therefore, there was no basis on the grounds of conservation to stop Dutch fishermen from fishing for North Sea plaice while fishermen of other nationalities could continue to do so. Romkes argued that quotas therefore discriminated between fishermen of different nationalities and was contrary to

Article 7 of the Treaty. The Dutch court asked for a ruling from the Court of Justice, which ruled as described.

It could therefore be claimed that the principle of relative stability has also been successfully implemented, at least in so far as the nine Member States which formed the Community in 1983 are concerned, if it were not for the fact that the application of Hague Preferences has affected some allocations. Greece, of course, does not fish in the area to which the system of TACs and quotas applies and has remained outside this controversy, although at one stage it did claim allocations. However, this was more to make a bargaining point over structures than a claim which it expected to be conceded.

Hague Preferences
As described in Chapter 4, Hague Preferences have not always been allocated in full, as apparently provided in the third recital quoted earlier on p. 118. This could be interpreted a failure of the policy. But the implementing article is Article 4 of Regulation No. 170/83 which provides that 'The volume of catches available to the Community . . . shall be distributed between the Member States in a manner which assures each Member State relative stability of fishing activities for each of the stocks considered.' The word 'relative' was deliberately used because the Commission realized that TACs would fluctuate from year to year and that absolute stability could not be guaranteed. If absolute stability could not be guaranteed, neither could specific Hague Preferences. If Hague Preferences had been allocated in full then the principle of distributing '. . . the catches available to the Community . . . on the basis of a reference allocation reflecting the orientations given by the Council . . .' could not have been observed.

For those stocks for which in some years the application of the allocation keys to the EC share of the TAC for certain stocks has resulted in quotas less than the Hague Preferences, the solutions were *ad hoc*, applicable for one year at a time. They did not represent a breakdown of the principle but its temporary modification to meet exceptional circumstances. In the case of stocks for which the recommended TACs were never large enough to respect the principles but which were fixed until 1991 at levels which enabled them to be respected, the question is more open. In fact, the Commission, in making its proposals for TACs and quotas for 1982, found itself in a situation which it had not foreseen.

In order to resolve this problem, the Commission made a political decision to respect the principles by ignoring the scientific advice, a decision justified at the time for the reasons explained in Chapter 3. The state of these stocks did not and is never likely to permit the principle of relative stability and the scientific advice to be respected simultaneously and they represent a special case. It is noteworthy in this context that the UK has never allocated the Hague Preferences which it obtained specifically to the regions on which they are based.

The 1992 review

The review provisions in Regulation No. 170/83 made only indirect reference to relative stability, requiring merely that the review of access and the 'Shetland Box' '. . . should take into account the objectives set out in Article 4(1) . . .', which were to assure '. . . each Member State relative stability of fishing activities for each of the stocks considered'. This placed no obligation on the Commission to propose any modifications. Nonetheless, the Commission not only did so but proposed sweeping changes to how the principle should be implemented. The Explanatory Memorandum to the proposal gave little hint of these, blandly stating that one of the objectives of the proposal was to maintain '. . . the principle of relative stability . . . without ruling out adjustments to some distribution keys, having due regard to the overall balance, in response to changes that have come about since the present distribution key was established'.

The preamble to the proposal showed otherwise, the relevant recitals stating, in somewhat opaque terms and infelicitous language:

> 'Whereas while observing the overall balance, it may become necessary to make adjustments to the distribution in the case of some availabilities to take account of developments with regard to biological and economic factors since 1983 and in particular miniquotas and certain traditional trade flows between Member States;
> Whereas, therefore, the objective of relative stability should be seen in those terms.'

(The term 'miniquotas' has never been explained but apparently refers to small quotas allocated to some Member States. The objective of these quotas was to provide for by-catches when fishing for other species, for which the Member State had large quotas, in order to minimize discarding. The quota of haddock allocated to Belgium, described earlier, fell into this category.)

The preamble describes the intentions of the legislation and thus acts as a guide to the courts in its interpretation. This preamble, if it had been adopted, would have completely altered the concept of relative stability.

Concerning the proposed legislation itself, the relevant paragraph revealed that the Commission proposed that '. . . shares in Community fishing opportunity shall be fairly divided among the Member States in the form of fishing availabilities expressed in terms of fishing effort and/or quotas allocated in such a way as to ensure relative stability of fishing activities . . .'. No indication was given as to how fishing effort was to be determined, which would be essential before it could be allocated. Furthermore, it was proposed that quotas should be allocated by the Management Committee procedure under which it would have been the Commission, not the Council, which would have allocated fishing availabilities to Member States and made any adjustments, to paraphrase the proposal.

This proposal raised howls of protest from most of the fishing industry. The concept of using fishing effort as a means of conservation had been ruled out in the negotiation of the conservation policy for the reason that it was impossible to quantify it in an acceptable manner. TACs and quotas were the core of relative stability. These proposals struck at the heart of the principle. In the event, little of the Commission's proposal survived into Regulation No. 3760/92. The new regulation does contain the provision for total allowable fishing effort to be fixed but the Council, acting on a proposal from the Commission, retains the responsibility for fixing allocations, whether of TACs or total allowable fishing effort. The wording on the distribution of fishing opportunities, whether by catches or fishing effort, remains almost identical in Regulation No. 3760/92 to that in Regulation No. 170/83; the legal basis for relative stability remains unchanged. The only concession to the Commission's proposal was that '. . . following a request from the Member States directly concerned, account may be taken of the development of miniquotas and regular quota swaps since 1983, with due regard to the balance of shares'. This leaves the initiative directly with the Member States concerned, without which the Commission can do nothing.

That the Commission considered it necessary to propose such sweeping changes means that it must have considered the application of the principle of relative stability unsatisfactory. An alternative view held by many, particularly in the UK fishing industry, was that it was a last-gasp attempt by a Spanish commissioner to destroy the principle of relative stability and open up fishing possibilities to Spain before he moved on to another portfolio. However, the Council conceded very little and retained almost exactly the same phraseology in Regulation No. 3760/92 as in Regulation No. 170/83. Most importantly, the wording of the relevant recitals remained identical. This was all important. Otherwise it would have provided the possibility for Spain and Portugal to take further cases to the Court of Justice. The review showed that the principle of relative stability has, for the majority of Member States, proved politically totally satisfactory. Only one Member State is wholly discontent. Spain voted against Regulation No. 3760/92.

The speed of adoption of this regulation provides another facet illustrating the importance to the Member States of retaining the principle of relative stability unaltered. The Commission did not publish its proposal until 6 October 1992, less than three months before the Council had to decide on any modifications. This in itself was surprising because it normally takes many months of careful negotiation to get controversial proposals adopted. It is possible that the Commission thought that it could use the lack of time to force its ideas through without proper discussion. Alternatively, the Commission had great difficulty in adopting the proposal because the commissioners were each fighting their national corners. Many of those concerned thought that the proposal would not be adopted by the deadline of 31 December 1992. That it was shows how much the Member States were determined to sink any differences in order to maintain what they consider such a vital principle – a further illustration of its political success.

Conclusions

The principle of relative stability has been successfully applied within the constraints placed upon it by the state of the stocks. Although this has created political problems, the compromise solutions which had to be found have been accepted and have not given rise to consequent problems. It may therefore be concluded that the principle of relative stability, although not an unqualified success, has been politically successful. This conclusion is supported by the fact that the provisions have been retained virtually unchanged in Regulation No. 3760/92.

The transfer of resources

It is not specifically stated in Regulation No. 170/83 that one of the objectives of the conservation policy was to transfer fish resources from some Member States to others, the provision being hidden by the phrase in the preamble '. . . on the basis of a reference allocation reflecting the orientations given by the Council'. The two orientations which involved transfer of resources were Hague Preferences, which benefited Ireland and the UK, and compensation for losses in third country waters, which benefited mainly Germany and the UK. That it was entitled '*compensation* for losses in third country waters' did not disguise the fact that resources had to be transferred to the Member States which benefited. These transfers have been achieved through the application of relative stability, which, as described in the previous section, has been applied as intended in most years.

The transfer of resources, a permanent but little recognized feature of the conservation policy, has therefore been successfully achieved, a further political success.

Socio-economic objectives

As already stated, Article 1 of Regulation No. 170/83 provides that the conservation policy shall be implemented '. . . in appropriate economic and social conditions . . .'. The implementation of this principle is very difficult, if not impossible to evaluate because nowhere is there any definition of what is meant by the phrase. Furthermore, the conservation policy is based only on scientific advice to which there is negligible economic and no social input. The only manner in which the achievement of this objective can be evaluated is to examine how the Council has operated in taking decisions.

Chapter 6 described the long delays in implementing scientific advice and the fact that when regulations were adopted they often did not reflect that advice or contained provisions which nullified its objectives. The fundamental reason for these actions of the Council was that to implement the advice as it stood would have had immediate, adverse socio-economic effects upon the fishing industry. This was self-evident. Given the over-exploited state of most fish stocks in Community waters, the predicted consequence of any proposed measure must be a reduction of catches in the short term. As already described in Chapter 6, pleading 'socio-economic' reasons for delaying or not adopting a proposal of the Commission was frequently used by Member States; no justification was ever offered or demanded by other Member States.

It is concluded that if '. . . appropriate social and economic conditions . . .' are those by which the Council decided to operate, the conservation policy must also be considered a political success in this respect.

A conservation policy?

Although the policy is termed a conservation policy, a fundamental question must be posed about all the principles which formed its political core, as examined above, which is 'Are any of them conservation measures?'

Restricting access to the coastal zone is not a conservation measure unless the intensity of fishing in the zone is limited. This did not happen. Although most Member States have national regulations limiting fishing in their coastal zones to small vessels, none has measures to restrict the total number of vessels. As shown by what happened in the continental coastal zone, described in Chapter 3, fishing intensity can build up to very high levels. Another example is the UK fisheries for lobsters and crabs which occur in this zone, the stocks of which are overexploited. So, in practice, access has not worked as a conservation measure, not surprisingly as this was never intended to be its objective; it was adopted as a social measure, to give preference to vessels operating from small coastal communities and protect them from the activities of larger vessels from other Member States.

Relative stability is also not a conservation measure and, in two important respects, has had adverse consequences for the conservation of the fish stocks. First, as already described, TACs have often been fixed at levels higher than the scientific advice in order that the application of this principle should not result in quotas less than the Hague Preferences. Second, in practice, each Member State often determines the size of the TAC for which it argues by first deciding the quota which it considers necessary to maintain the activities of its fleets and then dividing this by its percentage allocation, irrespective of the scientific advice. In the case of precautionary TACs, this is a standard method of operation. Member States justify this practice by arguing that in the absence of clear scientific advice that the stocks are endangered, quotas should not be fixed at levels which become limiting. Third, but most importantly, relative stability committed the Community to a system of conservation based upon TACs because these are the only practical means by which the principle can be implemented. As discussed in Chapter 8, for the majority of stocks TACs are a totally ineffective method of fisheries conservation as operated by the Community.

To summarize, the fundamental elements of the conservation policy have been implemented as intended to achieve their political objectives. In the review of the policy, the Council has not seen it necessary to alter or revise them in any way. On the contrary, the comparison between what the Commission proposed and what the Council adopted shows the firm commitment of the Member States to these principles. But these principles have no bearing upon conservation and their implementation has, in fact, been detrimental to the objectives of conservation. Almost certainly, the far-reaching proposals of the Commission were an attempt to lay the basis for improving this situation. The verdict must be that the conservation policy of the Community to date must be adjudged a brilliant political success but a conservation policy in name only.

Practical objectives

The criteria for success

Irrespective of whether the conservation policy is a conservation policy in name only, on 25 January 1983 the fishing industry definitely considered that the policy should achieve two major objectives which it considered the hallmarks of an effective conservation policy. These objectives were maintaining or increasing total catches and catch rates. There was also a third objective, which was stable catches, but the industry has an equivocal attitude to this. The processing sector considers this a primary objective because stable supplies are essential for it to make long-term plans for building processing plants and employing labour. The catching sector also regards stability as an objective but complaints about stability, or the lack of it, are usually voiced only when TACs are reduced from one year to the next, rarely if ever when they are increased. Implicit in the thinking of the fishing industry was that the achievement of these objectives would enable the number of fishing vessels and the number of men employed to increase and for individual vessels to become more powerful and more efficient. Above all else it meant greater prosperity. To what extent have these objectives been achieved?

Assessing the extent to which these objectives have been met is demanding, because it is difficult to decide which criteria to use and because, once having chosen the criteria, the necessary data are either not available or, if available, are not reliable.

Total allowable catches
TACs are the most definitive set of data available but they represent only catch possibilities, not actual total catches. For many stocks, especially those for which the TACs are precautionary, they do not even represent realistic catch possibilities. As described, TACs are often determined on the basis of fixing quotas which will not result in the fishery of any Member State being closed during the course of the year with the consequence that the TACs have been fixed at levels which exceed any recorded catches. Other TACs have been fixed on the basis of ensuring that the quotas of Member States are not less than its Hague Preference in that stock. Also, for those stocks for which allocations are made to third countries or the stocks are shared, the TACs do not represent the total catch possibilities available to Community fishermen. For some stocks the accession of Spain and Portugal has automatically increased the size of the TACs because the area and therefore the proportion of the stock to which the TAC applies has increased.

Total catches

Catch statistics should represent a better index of conservation success or failure. Unfortunately, official catch statistics are very unreliable for two reasons. First, 'catch statistics' are in fact 'landings statistics'. Second, fishermen misreport the area in which they caught their fish.

'Catches' are the quantities of fish taken out of the sea; 'landings' are the quantities landed and sold. The difference is 'discards'. Fishermen have always discarded fish for which there is no market but the system of TACs and quotas has resulted in two additional incentives to discard. First, fishermen have to discard fish for which their quota is exhausted. Second, fishermen keep only the larger and more valuable fish in order to maximize the value of their catches, a practice known as 'high-grading'. Quantities of discards are not officially recorded and, only in a few cases, even estimated. These estimates indicate that the difference between catches and landings may be as much as 30% or more in some cases.

Misreporting of the area of capture occurs when it is likely that a quota will be exhausted. Catches are then reported as coming from an area where the quota for the same species is unlikely to be exceeded or, in some cases, not fished at all. This is another consequence of the introduction of the system of TACs.

Because official statistics are so unreliable ACFM provides for most stocks estimates of catches which include 'unofficial landings' and discards, but these are not broken down by country. Total landings represent the best index of conservation success or failure but Community catches may be different for the same reason that the EC share of the TAC is less than the TAC. In any case, only official landings statistics are available by country.

In the following tables the estimates made by ACFM have been used for total catches. For those stocks for which all or most of the catches are taken by EC Member States the total catches, as estimated by ACFM, less the reported landings by third countries, have been used. In other cases the landings reported by Member States have been used.

Catch per unit effort

While total catches represent an overall index of conservation success or failure, for the individual fishermen, the best index is catch per unit effort or catch rate, the quantity of fish caught for each day of fishing at sea or hour's trawling or number of pots shot, whatever measure of fishing effort is chosen, because this represents the profits, the difference between earnings (= catches) and the cost of catching (= fishing effort). It is even more difficult to get a reliable index of catch per unit effort than of catches because the data on catches are unreliable, for the reasons already described and those on fishing effort are not available for many stocks. For those stocks for which indices of catch per unit effort are not available in the reports of the ICES Working Groups, indices of catch per unit effort have been derived by dividing the estimates of total catches by rates of fishing mortality; these are available only for those stocks for which analytical assessments are made.

Pelagic species

Seven pelagic species are subject to TACs: herring, sprat, anchovy, capelin, blue whiting, mackerel and horse mackerel. Of these, only the TACs for herring and mackerel are based upon analytical assessments and fixed with the objective of managing the stocks. Therefore, the data for these two species only are relevant to an evaluation of the success of the conservation policy.

Herring

The stocks of herring represent one of the best sets of data for evaluating the conservation policy because the majority of them are evaluated analytically and the scientific basis for the assessments is, in general, good. However, there are problems. In particular, the North Sea stock of herring is shared with Norway and until 1986 its management was not agreed.

In consequence, before this date, a TAC was fixed unilaterally by the Community. Now that the TAC is agreed by negotiation, the EC share is determined by the terms of the agreement between the Community and Norway which provides for the shares to vary with the size of the spawning stock biomass (see Table 4.1). Part of the EC share is then normally exchanged for other species, such as cod and haddock. However, the EC share has frequently been based upon a lower percentage than would be justified by the size of the spawning stock biomass in order to take account of EC over-fishing of its share in the previous year. Although this is never officially acknowledged, that it occurs is well known and obvious from the figures agreed during the negotiations.

The TACs and the EC share of those TACs fixed for all the stocks in Community waters are shown in Table 7.1. Those for the Baltic Sea and Skagerrak and Kattegat are not shown because these stocks are fished mainly by third countries and their state cannot be considered to be an index of the conservation policy; also the TACs for all the Baltic Sea stocks are fixed by the International Baltic Sea Fisheries Commission. The main features of this table are:

- the stock of North Sea herring (ICES divisions IVa, b, c and VIId), which was in a collapsed state in 1983, had recovered by 1985;
- since 1985 the EC share of this stock has never been less than 300 thousand tonnes except in 1990 and 1991 and then only marginally;
- the total EC share of the seven stocks has never fallen below 400 thousand tonnes and has been relatively stable over the period 1985–92;
- the TACs for the stocks to the west of Scotland and Ireland (VIa(S), VIb) and the Celtic Sea (ICES divisions VIIg, h, j, k) have more than doubled and that for the Irish Sea stock (ICES division VIIa) has almost doubled since 1982.

The data for estimated catches (Table 7.2) show that the total catches taken by Community fishermen increased by 117% from 1982 to 1990 and that for only two stocks, one of them of minor importance, were catches less in 1990 than in 1982. Comparison of Tables 7.1 and 7.2 show the extent to which estimated catches have exceeded the TACs and the EC shares of the TACs for all stocks in most years.

Table 7.1 TACs (T) and EC share (E) of stocks of herring (in thousands of tonnes).

Stock		1982	1983	1984	1985	1986	1987	1988	1989	1990	1991	1992	1993
TACs for stocks for which allocations made to third countries													
IVa, b	T	0	70	**	**	500	560	500	484	385	370	380	380
IVc, VIId	T	72	30*	55	90	70	40	30	30	30	50	50	50
VIa(N)	T	80	70	64.02	56.5	51.85	49.1	49.8	58	75	62	62	62
TACs or EC shares of the TACs													
IVa, b	E	0	41.06	100	228.65	298.425	334.4	294.85	312.675	260.15	245.35	250.85	250.85
IVc, VIId	E	68	0.03	55	90	70	40	30	30	30	50	50	50
VIa(N)	E	66.2	61.2	56	50.4	46.2	44.6	44.6	51.54	66.91	55.14	55.14	55.14
VIa(S)	T	11	12	12	14	17	17	14	20	27.5	27.5	28	28
Clyde	T	2.5	2.5	3	3	3.4	3.5	3.2	3.2	2.6	2.9	2.3	1.0
VIIa	T	3.8	3	5	5	6.3	4.5	10.5	6	7	6	7	7
VIIg, h, j, k	T	8.1	8.1	13	13	17.2	18	18	20	17.5	22	21	21
Totals		159.6	127.89	244	404.05	458.525	462	415.15	443.415	411.66	408.89	414.29	412.99

* This was not fixed by regulation; the situation was very confused because of fishing outside quotas before these had been finally decided.
** TAC not agreed with Norway.

Table 7.2 Herring: estimated EC total catches, in thousands of tonnes, and indices of catch per unit effort.

Stock	1982	1983	1984	1985	1986	1987	1988	1989	1990
Catches									
IV, VIId	121	92	153	297	301	320	440	439	361
VIa(N)*	79	55	68	39	77	59	47	53	68
VIa(S)*	19	33	27	23	29	49	29	29	44
Clyde	4	4	6	5	5	4	2	2	2
VIIa	5	4	4	9	7	6	10	5	6
VIIg, h, j, K*	16	33	28	27	33	37	30	32	28
Totals	244	221	286	400	452	475	558	560	530
Indices of catch per unit effort†									
IV, VIId	696	762	652	777	973	1183	1404	1550	1491
VIa(N)	201	163	167	156	190	203	206	279	280

* Includes estimates of discards.
† Based upon dividing total catches by fishing mortality rate.
Source : Report of ACFM for 1991.

Estimates of rates of fishing mortality for the period 1982–90 are available for the North Sea and west of Scotland (North) stocks only; for the former catch per unit effort has doubled over this period and for the latter increased by 40% (Table 7.2).

Mackerel

There are two stocks of mackerel in Community waters, the North Sea and western stocks. The former, jointly managed with Norway, had started to collapse by 1975 and the management regime has never created the conditions which might permit it to recover. The stock is jointly managed with Norway which has always insisted upon a TAC which permits a limited amount of fishing by communities based along the Norwegian coast; under these circumstances the Community has always insisted upon a share of the TAC for similar communities in the North Sea. The evolution of catches in this stock is therefore not a good index of the success or failure of the Community's conservation policy. Socio-economic considerations dictate not only the conservation policy of the Community but also that of Norway.

The management of the western mackerel stock is also complex because it does not occur exclusively in Community waters. In 1982 most catches were taken in Community waters but a change in its distribution resulted in an increasing proportion of the catches being taken in Norwegian waters and, to a lesser extent, in Faroese and international waters. Despite this, the Community insists that the stock is an autonomous stock for which the Community has exclusive competence to fix the TAC. Nonetheless, since 1989 the TAC has effectively been fixed during the annual negotiations with the Faroe Islands and Norway on fishing possibilities for the coming year. What happens is that each party agrees to fix an autonomous TAC if the other party does not exceed an agreed figure. Thus, the stock has been managed, even if management has been far from ideal.

Although the TAC has fluctuated, for most of the period it has never been less than 400 000 t (Table 7.3). After two years of very high catches in 1982 and 1983, when the Community fishery was virtually uncontrolled, estimated total catches, including discards, from the stock have fluctuated around 600 000 t (Table 7.3). This table also shows estimated catches, excluding discards, by the Community from this stock; in 1989 and 1990 there was agreement with Norway that catches by Community vessels in the northern part of the North Sea could be counted against the EC TAC for this stock and for this reason all catches by all Member States, except Denmark which had quotas in this area, have been allocated to this stock, as also catches by France in 1982–83 and by Scotland in 1987 when their catches were so high that it was improbable that they were taken from the North Sea stock of mackerel. Even though the catch statistics are unreliable, they show a stable fishery at high levels of catches.

Although catches have remained relatively stable, catch per unit effort has fallen by approximately 30% because the rate of fishing on this stock has increased by 40% since 1982.

Table 7.3 Western mackerel: recommended TACs, TACs, total estimated catches and EC landings, all in thousands of tonnes, and indices of catch per unit effort, 1982–90.

	1982	1983	1984	1985	1986	1987	1988	1989	1990	1991	1992
Recommended TAC	272	330	500	340	290	380	430	355	480	500	670
Actual TAC	401*	407*	443*	415*	367*	405*	573†	495†	525†	575†	423†
Total catches(1)	648	625	551	561	538	615	628	567	606	647	
EC landings(2)	459	428	418	362	313	369	352	334	354	352	
Catch per unit effort	295	298	278	267	245	267	224	227	216		

* EC TAC only.
† ICES sub-areas and divisions IIa, IVa, Vb, VI, VII, VIII excluding VIIIc, XII, XIV.
(1) Including estimated discards from 1984 onwards.
(2) See text for how catches have been estimated.
(3) The index of catch per unit effort is total catch divided by rate of fishing mortality.
Source: Reports of ACFM for 1991 and 1992.

Demersal species

Thirteen demersal species are subject to TACs: cod, haddock, saithe, pollack, Norway pout, whiting, hake, plaice, sole, megrim, monkfish, prawns and *Nephrops*. Of these, the TAC for prawns is a special case, being fixed only for those in the waters of French Guyana. None of the the TACs for pollack, Norway pout and *Nephrops* are based upon analytical assessments; those for pollack and *Nephrops* were introduced on the accession of Spain and Portugal in order to prevent free fishing on these species. This would have resulted in by-catches of other species subject to TAC for which neither Spain nor Portugal have quotas and which would have had to be discarded. The objective of the TAC for Norway pout is to limit by-catches of haddock and whiting. Of the other species, the TACs consist of a mixture of those based on analytical assessments and those which are precautionary. As for pelagic species and for the same reasons, the data for the stocks in the Baltic Sea and Skagerrak and Kattegat are not considered and in stating the number of stocks for which TACs are fixed for each species, those fixed for these areas are ignored.

Cod

TACs are fixed for four stocks, of which three, those for the North Sea, west of Scotland and the Irish Sea are based upon analytical assessments. As already stated, the TAC for Irish Sea cod was fixed every year until 1991 at a level which would result in the quota for Ireland respecting its Hague Preference. Even in 1991 it was fixed at 10 000 t, compared with the recommended TAC of 6000 t. Thus, no attempt has been made to manage this stock, leaving only two stocks relevant to this evaluation.

The data for the North Sea are given in Table 7.4: TACs, landings and catch per unit effort have all fallen since 1982, by more than 50%. For the west of Scotland cod stock (Table 7.5) TACs have also declined but only by 23%. For both stocks catch per unit effort has fallen in line with catches, rates of fishing mortality having remained constant at 60–65% a year.

Haddock

TACs are fixed for three stocks of haddock, of which one is entirely precautionary. A single TAC is fixed for the stock west of Scotland and Rockall, as for cod, but whereas almost no cod are caught at Rockall there exists a stock of haddock around Rockall. This is biologically separate from the stock west of Scotland and is assessed separately but the advice on catch possibilities is not usually based upon an analytical assessment. This makes comparison between the recommended TACs and those fixed difficult.

For North Sea haddock (Table 7.6), TACs and catches have fallen considerably since 1982, the TAC for 1992 being only one third of that for 1982. Landings have fallen by 75% but estimated total catches by only 62%. As for cod, catch per unit effort reflects catches; for this stock, rates of fishing mortality have also remained almost constant since 1982 but at an even higher level than those for North Sea cod, the rate of removal by fishing being of the order of 65% a year.

Table 7.4 North Sea cod: recommended TACs, TACs, and total and EC landings, all in thousands of tonnes, and indices of catch per unit effort, 1982–91.

	1982	1983	1984	1985	1986	1987	1988	1989	1990	1991	1992
Recommended TAC	<235	<220	<182	<259	<130	<125	<148	<124	113	92	82
Actual TAC	235	240	215	250	170	175	160	124	105	100	101
Total landings	256	237	197	188	157	167	142	111	105	84	
EC landings	243	230	190	181	152	161	138	105	99(4)	78	
Catch per unit effort index	303	262	241	235	204	197	170	118	132	90	

Notes:
(1) For the period 1982–89, ACFM recommended that TACs should be fixed at rates of fishing mortality which corresponded with a rate less than the existing rate; the TAC given is that corresponding to no reduction.
(2) For 1991 and 1992 ACFM recommended that fishing effort should be reduced by 30%; the TACs given are those corresponding to a reduction of 30% in the rate of fishing mortality.
(3) The index of catch per unit effort is total catch divided by rate of fishing mortality.
(4) Excluding France for which no catch data available.

Source: Report of ACFM for 1991.

Table 7.5 West of Scotland cod: recommended TACs, TACs, EC catches, all in thousands of tonnes, and indices of catch per unit effort.

	1982	1983	1984	1985	1986	1987	1988	1989	1990	1991	1992
Recommended TAC	18	26.3	23.5	27.5	25.5	22.5	16.5	16.5	16	14.5	12
Actual TAC	17.5	27	25	25	25	22	18.4	18.4	16	16	13.5
EC landings	22	22	21	19	13	20	20	NA	NA	16	
Catch per unit effort index	30	26	24	19	14	19	22				

Notes:
(1) For the period 1982–89, ACFM recommended that TACs should be fixed at rates of fishing mortality which corresponded with a rate less than the existing rate; the TAC given is that corresponding to no reduction.
(2) For 1991 and 1992 ACFM recommended that fishing effort should be reduced by 30%; the TACs given are those corresponding to a reduction of 30% in the rate of fishing mortality.
(3) Most of the catches are taken by Member States.
(4) The index of catch per unit effort is catch divided by rate of fishing mortality.
NA = not available.

Source: Report of ACFM for 1992.

Table 7.6 North Sea haddock: recommended TACs, TACs, estimated total catches and EC landings, all in thousands of tonnes, and indices of catch per unit effort, 1982–90.

	1982	1983	1984	1985	1986	1987	1988	1989	1990	1991	1992
Recommended TAC	200	170	<162	<209	<239	<120	<185	<68	50	48	61
Actual TAC	180	181	170	207	230	140	185	68	50	50	60
Total catches	NA	238	213	251	220	172	171	104	87	90	NA
EC landings	169	159	128	163	160	106	102	62	42†	41	
Catch per unit effort index	289	179	215	267	206	169	163	114	94	92	

Notes:

(1) For the period 1985–89, ACFM recommended that TACs should be fixed at rates of fishing mortality which corresponded with a rate less than the existing rate; the TAC given is that corresponding to no reduction.

(2) For 1991 and 1992, ACFM recommended that fishing effort should be reduced by 30%; the TACs given are those corresponding to a reduction of 30% in the rate of fishing mortality.

(3) Total catches as estimated by ACFM.

(4) The index of catch per unit effort is total catch as estimated by ACFM, including discards, divided by rate of fishing mortality.

† Provisional; data for some countries not available.

NA = not available

Source: Reports of ACFM for 1990 and 1992.

For the west of Scotland and Rockall stock the situation is similar (Table 7.7). TACs and catches have declined since 1982 but not so markedly as for the North Sea; catch per unit effort for the west of Scotland stock has declined considerably because not only have catches fallen but the rate of fishing mortality has increased from 40% a year in 1982–83 to 60% a year.

Saithe

The TACs for two of the three stocks for which TACs are fixed are based upon analytical assessments. As for cod and haddock, the North Sea stock is jointly managed with Norway but unlike cod and haddock, Norway has the major interest in the saithe stock and largely determines the size of the TAC. Its management is therefore not a good index of the success of the Community's conservation policy but it is examined, nevertheless. TACs rose from 1982 to 1986 and then fell, being lower in 1992 than in 1982.

Catches also increased from 1982 onwards, reaching a maximum in 1985 since when they have also fallen but to a much larger extent than the TACs (Table 7.8). Catch per unit effort has fallen to 46% of its 1982–83 level.

For the west of Scotland and Rockall stock TACs for the years 1985–87 were fixed at a level which ensured that the UK Hague Preference was respected (Table 7.9). Apart from this, the pattern of TACs and catches is similar to the North Sea stock except that catch per unit effort, after increasing in the period 1984–88, has fallen.

Whiting

Of the five stocks of whiting for which TACs are fixed, three are based upon analytical assessment. However, as already described, the TACs for the west of Scotland and Irish Sea stocks were fixed for the period 1982–89 at levels which far exceeded those advised by ACFM in order to respect the UK Hague Preferences. Consequently, only the data for North Sea whiting provide a basis for evaluation. However, evaluation is complicated for the reasons that part of the catches is taken as a by-catch in the industrial fishery for Norway pout and a high proportion of the catches taken in the human consumption fishery is discarded. Account is taken of both by-catches and discards in making the assessments and fixing TACs but it complicates the evaluation.

The recommended maximum TACs have remained relatively stable since 1982, in most years being in the range 120 000–135 000 t (Table 7.10). In 1982 and 1983, TACs were fixed much higher than those recommended by ACFM in order to achieve percentage allocations which permitted the settlement of the conservation policy, as described in Chapter 3. Since then the TACs fixed have largely corresponded with the maximum level advised by ACFM, except for 1984. Total catches have decreased by 23% since 1983 but landings in the human consumption fisheries have fallen, to only 40% of their 1982–83 level in 1989. Catch per unit effort in the human consumption fishery has fluctuated without apparent trend.

Table 7.7 West Scotland and Rockall haddock: recommended TACs, TACs and estimated EC catches, all in thousands of tonnes, and indices of catch per unit effort, 1982–90.

	1982	1983	1984	1985	1986	1987	1988	1989	1990	1991	1992
Recommended TAC	21.5	44.4	<47	<33	—	<33	35	33	19.5	13	10.8
Actual TAC	21.5	45	40	36	34.5	32	35	35	24	15.2	12.5
EC catches	34	31	30	33	25	35	29	NA	NA		
Catch per unit effort index	65	59	39	38	45	29	28				

Notes:
(1) For the years 1984, 1985 and 1987, ACFM recommended that TACs should be fixed at rates of fishing mortality which corresponded with a rate less than the existing rate; the TAC given is that corresponding to no reduction; a specific TAC was not recommended for west of Scotland in 1986.
(2) For 1991 and 1992, ACFM recommended that fishing effort should be reduced by 30%; the TACs given are those corresponding to a reduction of 30% in the rate of fishing mortality.
(3) The index of catch per unit effort is total catch as estimated by ACFM, including discards, divided by rate of fishing mortality. Catch per unit effort index is based upon data for west of Scotland stock only.
NA: data for some countries not available.
Source: Report of ACFM for 1992.

Table 7.8 North Sea saithe: recommended TACs, TACs and estimated total catches and EC landings, all in thousands of tonnes, and indices of catch per unit effort, 1982–90.

	1982	1983	1984	1985	1986	1987	1988	1989	1990	1991	1992
Recommended TAC	100	131	160	<195	195	<198	<156	170	120	125	102
Actual TAC	125	158	180	200	240	173	165	170	120	125	110
Total catches	NA	169	198	194	167	154	113	93	NA	NA	
EC landings	84	76	93	89	97	84	69	65	NA	NA	
Catch per unit effort index	319	284	261	235	176	216	162	128	135	132	

Notes:
(1) For the years 1985, 1987 and 1988, ACFM recommended that TACs should be fixed at rates of fishing mortality which corresponded with a rate less than the existing rate; the TAC given is that corresponding to no reduction.
(2) Total catches as estimated by ACFM.
(3) The index of catch per unit effort is total catch as estimated by ACFM, including discards, divided by rate of fishing mortality.
NA: data for some countries not available.
Source: Reports of ACFM for 1990 and 1992.

Table 7.9 West of Scotland and Rockall saithe: recommended TACs, TACs and estimated total catches, all in thousands of tonnes, and indices of catch per unit effort, 1982–90.

	1982	1983	1984	1985	1986	1987	1988	1989	1990	1991	1992
Recommended TAC	13	8.2	27	26	20	23	35	20	24	21	19
Actual TAC	13	16	27	27.8	27.8	27.8	35	30	29	22	17
Total catches(1)	22	26	27	26	35	33	33	NA	NA		
Catch per unit effort index	39	50	92	96	70	66	65	34	31	27	

Notes:
(1) Most of the catch is taken by EC Member States.
(2) The index of catch per unit effort is total catch as estimated by ACFM, including discards, divided by rate of fishing mortality. NA Data for some countries not available.

Source: Reports of ACFM for 1991 and 1992.

Table 7.10 North Sea whiting: recommended TACs, TACs, estimated total catches and EC landings, all in thousands of tonnes, and indices of catch per unit effort, 1982–90.

	1982	1983	1984	1985	1986	1987	1988	1989	1990	1991	1992
Recommended TAC	200	125	<86	<118	<135	<127	<134	<115	130	134	120
Actual TAC	170	170	149	160	135	135	120	115	125	141	135
Total catches (3)	NA	151	135	97	154	132	127	118	147	117	
Total landings (4)	100	99	99	72	67	65	66	40	NA	46	
Catch per unit effort index	161	149	155	118	173	120	155	146	169*	122	

Notes:

(1) For the period 1984–89, ACFM recommended that TACs should be fixed at rates of fishing mortality which corresponded with a rate less than the existing rate; the TAC given is that corresponding to no reduction.

(2) For 1991 and 1992 ACFM recommended that fishing effort should be reduced by 30%; the TACs given are those corresponding to a reduction of 30% in the rate of fishing mortality.

(3) Including estimated discards and by-catches in the fishery for Norway pout.

(4) Catches in the human consumption fisheries; most of the catches are taken by Member States.

(5) The index of catch per unit effort is total catch as estimated by ACFM, including discards, divided by rate of fishing mortality. NA Data for some countries not available.

Source: Reports of ACFM for 1990 and 1992.

Plaice
TACs are fixed for eight stocks of plaice, of which only two, those for the North Sea and Irish Sea, have been based upon analytical assessments for the whole of the period 1982–92. For the North Sea stock TACs have fluctuated between 140 000–200 000 t (Table 7.11). However, estimated catches have risen, being 10% higher in the period 1988–90 than in the period 1982–84. The data for catch per unit effort are contradictory, those for Belgian and Danish vessels showing a slight increase compared with a marked decrease for UK vessels.

For the Irish Sea stock, TACs increased from 1982 to 1989 since when they have fallen to less than their 1982 level. Catches have shown a similar trend (Table 7.12). In contrast, catch per unit effort more than doubled between 1982 and 1989.

For the five most important plaice stocks in Community waters, TACs have been stable throughout the period 1983–92 (Table 7.13).

Sole
TACs are fixed for ten stocks of soles, of which data for the period 1982–92, based upon analytical assessments, are available for four.

For the stock of North Sea sole TACs were 26% higher in the period 1990–92 than in the period 1982–84, although in the period 1987–89 they fell considerably. Total catches show a similar trend (Table 7.14). However, catch per unit effort in 1989 was only 88% of that in 1982.

For the Irish Sea stock of sole (Table 7.15) both TACs and catches increased from 1982 to 1987, since when they have fallen, although the TAC for 1992 was higher than that for the period 1982–85; catch per unit effort rose from 1983 to 1989.

For the stock of sole in the eastern English Channel (Table 7.16) TACs and total catches increased continuously from 1982 to 1989. The TAC for 1992 was less than that for 1991 although still 35% higher than that fixed for 1982. Catch per unit effort, after increasing from 1982 to 1986, declined and was lower in 1990 than in 1982. The stock in the western English Channel (Table 7.17) shows a different picture, TACs, total catches and catch per unit effort all having fallen since 1982, TACs by 53%, catch per unit effort by 58% and total catches by 50%.

Considering all the main stocks together, TACs were 28% higher in the period 1991–92 than in 1982–83 (Table 7.18).

Hake, megrim and monkfish
The TACs for these stocks are not based upon analytical assessments but, being important stocks, are included to show that, for most stocks, they increased from 1982 to 1992 (Table 7.19).

Table 7.11 North Sea plaice: recommended TACs, TACs and estimated total catches, and indices of catch per unit effort, 1982–90.

	1982	1983	1984	1985	1986	1987	1988	1989	1990	1991	1992
Recommended TAC	140	164	150	130	<160	120	150	<175	171	169	135
Actual TAC	140	164	182	200	180	150	175	185	180	175	175
Total catches (3)	155	144	156	163	165	153	162	170	168	154	
Catch per (4)	63.5	70.1	67.4	61.4	56.1	66.0	71.9	74.6			
unit (5)	98.7	60.4	52.7	42.2	48.6	59.0	58.4	53.2			
effort (6)	—	—	626	605	599	695	603	734	767		

Notes:

(1) <: ACFM recommended that TACs should be fixed at rates of fishing mortality which corresponded with a rate less than the existing rate; the TAC given is that corresponding to no reduction.

(2) For 1992 ACFM recommended that fishing effort should be reduced by 20%; the TAC given is that corresponding to a reduction of 20% in the rate of fishing mortality.

(3) As estimated by ACFM; in all years there have been illegal landings varying from an estimated 40–160 thousand tonnes.

(4) Index based upon hours fishing by Belgian beam trawlers, corrected for horsepower.

(5) kg/h, UK beam trawlers.

(6) kg/dy by Danish seiners (Denmark).

Sources: Reports of ACFM for 1991 and 1992 except catch per unit effort data, North Sea Flatfish Working Group Report for 1991, CM 1991/Assess:5.

Table 7.12 Irish Sea plaice: recommended TACs, TACs and estimated total catches, all in thousands of tonnes, and indices of catch per unit effort, 1982–90.

	1982	1983	1984	1985	1986	1987	1988	1989	1990	1991	1992
Recommended TAC	3.0	3.5	3.1	4.0	5.0	5.0	4.8	5.8	5.1	3.3	3.0
Actual TAC	4.5	4.5	4.5	5.0	5.0	5.0	5.0	5.8	5.1	4.5	3.8
Total catches (1)	3.2	3.6	4.2	5.1	4.8	6.2	5.0	4.4	3.3	2.5	
Catch per unit effort (2)	3.5	6.2	5.3	7.1	7.8	7.3	5.4	7.4			

Notes:
(1) Estimated by ACFM.
(2) kg/h, Belgian beam trawlers.

Sources: Reports of ACFM for 1991 and 1992 except catch per unit effort data, Report of the Irish Sea and Bristol Channel Working Group Report for 1990, ICES CM. 1991/Assess:1.

Table 7.13 TACs for the main stocks of plaice in Community waters, 1982–93 (thousands of tonnes).

Stock	1982	1983	1984	1985	1986	1987	1988	1989	1990	1991	1992	1993
IV	140	164	182	200	180	150	175	185	180	175	175	175
VIIa	4.5	4.5	4.5	5	5	5	5	5.8	5.1	4.5	3.8	2.8
VIId,e	5.5	6.5	6	6.5	6.9	8.3	9.96	11.7	10.7	10.7	9.6	8.5
VIIf,g	1.45	1.2	1.4	1.4	1.8	2	2.5	2.5	1.9	1.9	1.5	1.4
VIIh,j,k	0.8	0.8	0.8	0.8	0.8	0.96	1.15	1.15	1.15	1.35	1.35	1.35
Totals	152.25	187	194.7	213.7	194.5	166.26	193.61	206.15	198.85	193.45	191.25	189.05

Table 7.14 North Sea sole: recommended TACs, TACs and estimated total catches, all in thousands of tonnes, and indices of catch per unit effort, 1982–90.

	1982	1983	1984	1985	1986	1987	1988	1989	1990	1991	1992
Recommended TAC	15	15	14	15	12	11	11	14	25	27	21 (1)
Actual TAC	21	20	20	22	20	14	14	14	25	27	25
Total catches (2)	22	25	27	24	18	17	22	22	35	38	
Catch per unit effort (3)	310	320	307	276	213	205	236	273			

Notes:
(1) For 1992 ACFM recommended that TACs should be reduced by 20%; the recommended TAC shown is that corresponding to a reduction of 20% in the rate of fishing mortality.
(2) As estimated by ACFM; in all years there have been illegal landings varying from an estimated 40–160 thousand tonnes.
(3) Index based on (horsepower × days at sea) by Dutch beam trawlers.

Sources: Reports of ACFM for 1991 and 1992 except catch per unit effort data, North Sea Flatfish Working Group Report for 1991, CM 1991/Assess:5.

Table 7.15 Irish Sea sole: recommended TACs, TACs and estimated total catches, all in thousands of tonnes, and indices of catch per unit effort, 1982–90.

	1982	1983	1984	1985	1986	1987	1988	1989	1990	1991	1992
Recommended TAC	1.6	0.7	1.0	1.1	1.65	1.9	1.6	<1.48	1.5	1.3	1.24
Actual TAC	1.25	1.25	1.25	1.25	1.9	2.1	1.75	1.48	1.5	1.5	1.35
Total catches (2)	1.3	1.2	1.1	1.1	2.0	2.8	2.0	1.8	1.6	1.2	
Catch per unit effort (3)	16.2	12.8	14.3	14.6	12.9	12.6	14.7	18.2			

Notes:
(1) <: ACFM recommended that TACs should be fixed at rates of fishing mortality which corresponded with a rate less than the existing rate; the TAC shown is that corresponding to no reduction.

(2) As estimated by ACFM.

(3) kg/h, Belgian beam trawlers, second quarter of the year.

Sources: Reports of ACFM for 1991 and 1992 except catch per unit effort data, Irish Sea and Bristol Channel Working Group Report for 1990, ICES CM 1991/Assess:1.

Table 7.16 Eastern English Channel sole: recommended TACs, TACs and estimated total catches, all in thousands of tonnes, and indices of catch per unit effort, 1982–90.

	1982	1983	1984	1985	1986	1987	1988	1989	1990	1991	1992
Recommended TAC	—	2.4	1.4	2.3	2.6	3.1	3.4	3.8	3.7	3.4	2.7
Actual TAC	2.6	2.1	2.5	2.7	3.2	3.85	3.85	3.85	3.85	3.85	3.5
Total catches (1)	2.8	3.2	3.3	3.9	3.9	4.9	3.9	4.2	4.0	4.3	
Catch per unit effort (2)	11.9	11.2	10.4	10.3	14.7	13.3	8.4	9.9			

Notes:
(1) As estimated by ACFM.
(2) kg/h, Belgian beam trawlers.
Sources: Reports of ACFM for 1991 and 1992 except catch per unit effort data, Report of the North Sea Flatfish Working Group for 1990, ICES CM 1991/Assess:5.

Table 7.17 Western English Channel sole: recommended TACs, TACs and estimated total catches, all in thousands of tonnes, and indices of catch per unit effort, 1982–90.

	1982	1983	1984	1985	1986	1987	1988	1989	1990	1991	1992
Recommended TAC	1.7	0.4	0.9	1.3	1.3	1.3	1.3	1.0	0.9	0.54	0.77
Actual TAC	1.7	1.1	1.35	1.4	1.3	1.15	1.3	1.0	0.9	0.8	0.8
Total catches (1)	1.4	1.5	1.4	1.4	1.4	1.2	1.4	1.2	1.1	0.7	
Catch per unit effort (2)	10.9	9.0	9.2	8.6	9.7	7.6	7.5	4.6			

Notes:
(1) As estimated by ACFM.
(2) kg/h, UK beam trawlers.

Sources: Reports of ACFM for 1991 and 1992 except catch per unit effort data, Report of the North Sea Flatfish Working Group for 1990, ICES CM 1991/Assess:5.

Table 7.18 TACs, in thousands of tonnes, for the main stocks of sole in Community waters.

Stock	1982	1983	1984	1985	1986	1987	1988	1989	1990	1991	1992	1993
Year												
IV	21	20	20	22	20	14	14	14	25	27	25	32
VIIa	1.6	1.35	1.25	1.25	1.9	2.1	1.75	1.48	1.5	1.5	1.35	1.0
VIId	2.6	2.1	2.5	2.7	3.2	3.85	3.85	3.85	3.85	3.85	3.5	3.2
VIIe	1.7	1.1	1.35	1.4	1.3	1.15	1.3	1.0	0.9	0.8	0.8	0.9
VIIf,g	1.6	1.44	1.2	1.2	1.5	1.6	1.1	1.0	1.2	1.2	1.2	1.1
VIIh,j,k	—	0.6	0.6	0.6	0.6	0.6	0.6	0.72	0.72	0.72	0.72	0.72
VIIIa,b	3.1	3.1	3.1	3.305	3.4	4.44	4.0	4.8	5.2	5.3	5.3	5.7
Totals	31.6	29.69	30.0	32.455	31.9	27.74	26.6	26.85	38.37	40.37	37.87	44.62

Table 7.19 TACs, in thousands of tonnes, for hake, megrim and monkfish.

Stock	1982	1983	1984	1985	1986	1987	1988	1989	1990	1991	1992	1993
Hake												
VI, VII	35.75	20.63	19.91	20	25.19	36	37.5	33.75	36.89	37.59	38.75	40.2
VIIIa, b, d, e	†	15.63	15.15	15.3	19.81	24	25	22.5	24.6	25.06	25.83	26.8
VIIIc, IX, etc.	—	—	—	—	29.3	25	25	20	20	18	16	12
Megrim												
VI	—	11	17	4.4	4.4	4.4	4.84	4.84	4.84	4.84	4.84	4.84
VII	—	†	†	11.98	14.44	14.44	15.88	15.88	15.88	15.88	15.88	19
VIIIa, b, d, e	—	†	†	1.12	2.02	2.02	2.22	2.22	2.22	2.22	2.22	2.46
VIIc, IX	—	—	—	—	13	13	13	13	13	14.3	14.3	8
Monkfish												
VI	—	32.5	41.5	7.8	7.82	7.82	8.6	8.6	8.6	8.6	8.6	8.6
VII	—	†	†	25.87	30.07	30.07	33.08	33.08	33.08	33.08	33.08	19.24
VIIIa,b,d	—	†	†	8.33	9.01	9.01	9.91	9.91	9.91	9.91	9.91	5.76
VIIIc, IX	—	—	—	—	12	12	12	12	12	12	12	13

† Single TAC for VI, VII, VIII (EC zone).

North Sea cod and haddock: mismanagement or environmental change?

The recent history of the North Sea cod and haddock stocks has been categorized by the British fishing industry as demonstrating the total failure of the CFP, meaning the conservation policy, and led to calls for its abandonment. In the context of evaluating the conservation policy this history therefore needs to be examined in detail.

Tables 7.4 and 7.6 show that both the TACs and catches of North Sea cod and haddock have fallen very considerably since the adoption of the conservation policy. The decline in the abundance of the two stocks led ACFM in 1990 to express its serious concern about the low levels of the spawning stock biomasses and to refuse to provide catch possibilities, recommending instead that fishing effort should be reduced by 30%. This caused the Commission to propose corresponding TACs, although only for haddock in 1991 did the TAC adopted by the Council correspond with such a reduction. In addition, both for 1991 and 1992, regulations were adopted limiting the number of days at sea by vessels fishing primarily for these two stocks, as described in Chapter 6. The British fishing industry, which is heavily dependent upon these two stocks, having 47% of the EC share of the TAC for cod and 78% of that for haddock described the situation as a crisis for the CFP. It certainly was a crisis for the CFP but was it provoked by the operation of the conservation policy or would it have happened if the policy had not existed?

Statistics of landings from the North Sea of cod and haddock are published in the Bulletins Statistiques of ICES. Those for the period 1920 to 1986 are summarized in Table 7.20. The periods of the two major European wars and the years immediately following have been excluded, the former because fishing was very restricted and the latter because catches were high for the reason that the stocks increased during the war because there was little fishing. During the period 1920–38 and 1947–63, the maximum recorded annual landings of cod were 129 000 t in 1920 and the minimum 52 470 t in 1936. Average landings were slightly less than 81 000 t with little difference between the two periods 1920–38 and 1947–63. Landings of haddock fluctuated more than those of cod, the maximum landings being 210 019 t in 1920 and the minimum 52 419 t in 1962. Over the whole period considered, landings averaged just over 95 000 t, those in the period 1920–38 being higher than those in 1947–63. From 1964 onwards landings of both species started to increase dramatically. For the period 1964–86 average annual landings of cod were almost three times the average for 1920–38 and 1947–63 and those of haddock more than twice. The most remarkable feature was that levels of landings which had hitherto been the maxima became for haddock the average and for cod nearly twice the average. This phenomenon was termed 'the gadoid outburst' (cod and haddock belong to the scientific family, *Gadidae*). It was clearly recognized at the time that the increase in landings was a direct consequence of an increase in recruit-ment, the specific reason for which has never been determined. The increase in recruitment was most marked for cod (Table 7.21), for which average recruitment in the

Table 7.20 Average landings, in thousands of tonnes, of North Sea cod and haddock, 1920–86.

Period	Cod	Haddock
1920–38	74.6	116.1
1947–63	87.5	72.0
1920–38 + 1947–63	80.7	95.3
1964–86	227.9	213.8

Source: ICES Bulletins Statistiques.

period 1963–85 was 163% higher than in the period 1954–62; the start of the increase can be dated almost exactly to 1963. Even though recruitment fell after 1985, it was still 47% higher in the period 1986–90 than in the period 1954–62. For haddock, there were two very large year-classes in 1961 and 1962 followed by a decade, 1965–74, of sustained high recruitment which included the largest year-class on record, that of 1967. Subsequently, recruitment fell to an all-time recorded low level, which was only 31% of that in the period 1965–74 (Table 7.22).

Tables 7.21 and 7.22 show that the decline in landings of cod and haddock since the adoption of the conservation policy in 1983 occurred as a result of very marked changes in recruitment. In 1963 environmental conditions must have changed in a manner which favoured the survival of young cod and haddock, resulting in high recruitment throughout the late 1960s, 1970s and early 1980s. The environment has now apparently reverted to what, in historic terms, may be termed 'normal' as also have catches (Fig. 7.1). Even though recruitment to the haddock, but not the cod, stock improved in 1991 and 1992, resulting in an increase in the TAC from 60 000 t in 1992 to 133 000 t in 1993, this does not invalidate the conclusion to be drawn from these events. The decline in landings was a consequence of recruitment falling from its previous very high levels; it was not a failure of the conservation policy.

The adoption of the conservation policy coincided almost exactly with the end of the gadoid outburst. This very unfortunate coincidence has led to a totally erroneous appreciation of the policy by fishermen, especially those in the UK who were worst affected. It is not surprising that fishermen should have failed to appreciate the true reason for the decline in landings as the majority would have spent most, if not all their time at sea during the period of high catches and would have had few memories, if any, of the earlier period. If the gadoid outburst had started in 1983 instead of 1963 the CFP might now be regarded by the British fishing industry as a major success. On the other hand, the cause might have been recognized, as was the case with English Channel cod, for what it was, an increase in recruitment. However, in the case of English Channel cod, accepting the scientific advice enabled the

Table 7.21 Estimates of recruitment for North Sea cod, 1954–91, in millions of one-year-olds.

Year-class	Number	Year-class	Number	Year-class	Number
1954	87	1967	169	1980	271
1955	267	1968	190	1981	557
1956	74	1969	729	1982	269
1957	183	1970	847	1983	534
1958	310	1971	159	1984	108
1959	69	1972	289	1985	581
1960	69	1973	232	1986	257
1961	196	1974	426	1987	201
1962	114	1975	196	1988	324
1963	332	1976	726	1989	136
1964	382	1977	426	1990	155
1965	451	1978	449	1991	(341)
1966	50	1979	800		

Data for 1991 are provisional

Averages: 1954–62 152
1963–85 399
1986–91 236

Sources:
1954–62: Holden, 1978 (catch per unit effort of 2-year-old cod).
1963–68: Report of the North Sea Working Group, 1985 (ICES C.M. Assess:9): millions of one-year-old cod, interpolated from Figure 5.1B.
1969–88: Report of the Roundfish Working Group 1991 (ICES C.M. Assess:4): millions of one-year-old cod.
1989–91: Report of ACFM for 1992 (millions of one-year-old cod).

Notes:
There is not a continuous series of recruitment data for the period 1954–90; also the methods of calculation have changed over the period 1963–90. The series was derived using raising factors based upon comparisons of two sets of data, for which the years of observations overlapped, in order to obtain comparative data in standard units.

Data for 1954–62 were raised using a factor of 1.16, based upon the average of the ratios of the data for 1963–71, and then raised by a factor 1.92, based upon the average of the ratios of the data for 1969–84.

Data for 1963–68 were raised by the factor of 1.92.

industry to argue for higher TACs. In the case of North Sea cod and haddock, ignoring the scientific advice provided the industry with a stick with which to beat the donkey of the CFP which they detest.

But it is not entirely fair to blame the industry for their appreciation of what happened. The scientific advice on this issue was presented in a misleading manner because the fishery scientists unfortunately fell into a trap. The introduction of the mathematical model which permits catch possibilities to be calculated coincided with the start of the gadoid outburst; this model also provides absolute estimates of recruitment and spawning stock biomasses. Once fishery scientists had these estimates, they ignored the earlier data based upon recruitment surveys. In consequence, the accepted baseline for recruitment was set at the very high, one might say abnormally high, levels pertaining in the period mid-1960s to 1980. The same occurred for spawning stock

Table 7.22 Estimates of recruitment for North Sea haddock, 1925–92, in billions of 0-year-olds.

Year-class	Number	Year-class	Number	Year-class	Number
1925	76	1951	162	1972	20
1926	123	1952	97	1973	67
1927	25	1953	94	1974	122
1928	279	1954	102	1975	10
1929	52	1955	140	1976	15
1930	24	1956	6	1977	25
1931	146	1957	4	1978	37
1932	25	1958	90	1979	67
1933	56	1959	28	1980	15
1934	20	1960	18	1981	30
1935	166	1961	102	1982	19
1936	21	1962	128	1983	63
1937	16	1963	2	1984	16
1938	54	1964	14	1985	23
1939–44	No data	1965	42	1986	45
1945	148	1966	80	1987	6
1946	2	1967	259	1988	11
1947	16	1968	12	1989	13
1948	4	1969	14	1990	34
1949	32	1970	80	1991	67
1950	26	1971	74	1992	(57)

Data for 1990 are provisional

Averages:	1925–29	111	1960–64	53
	1930–34	54	1965–69	81
	1935–38	64	1970–74	73
	1945–49	40	1975–79	31
	1950–54	96	1980–84	20
	1955–59	54	1985–89	20
			1990–92	53

Sources:
1925–59: Jones and Hislop, 1978; (Table 23: catch per unit effort of one-year-old haddock).

Sources for other data as Table 7.21 except data for 1960–69 interpolated from Figure 9.1B of 1985 ICES C.M. Assess: 9 and data for 1970–89 from Table 18.10 of ICES C.M. Assess:4.

Note:
The same method of constructing a continuous series was used as for Table 7.21; data for 1925–59 were raised using a factor of 41.96, based upon the average of the ratios of the data for 1960–72, and then subsequently raised by a factor 18.94, based upon the average of the ratios of the data for 1960–84.

Data for 1960–69 were raised by the factor of 18.94.

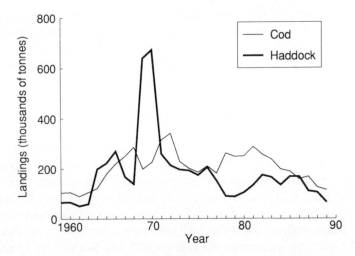

Fig. 7.1 Landings of North Sea cod and haddock, 1960–89, illustrating how the level of landings in the late 1980s was similar to that in the early 1960s.

biomasses. ACFM has established 'the lowest desirable' level for cod as 150 000 t and for haddock 100 000 t, based upon data since 1963.

Estimates of the sizes of the spawning stock biomasses prior to 1963 can be obtained by considering catches and rates of fishing mortality. The latter provides an estimate of the proportion of the total stock which is removed by fishing each year so, for example, if 50% of the stock is removed in a year and catches are 50 000 t, the size of the total stock must be of the order of 100 000 t; the calculation is slightly more complicated than this because account must be taken of deaths from natural mortality but the simplified calculation demonstrates the principle. For haddock, data on rates of fishing mortality are available since 1926; these show that rates have varied very little over the whole period 1926–92, being of the order of 45–55% a year, or even higher. For cod, the data are not available for the 1920s but the rate of fishing was about 50% in the 1930s, fell to about 45% in the 1960s and has been 45–50% in the 1980s. Because the rates of removal have been relatively unchanged over the whole period, the catches shown in Table 7.20 are proportional to the total exploitable stock biomass and, assuming that the same age groups have been exploited in the same manner throughout this period, also to spawning stock biomasses. These figures indicate that throughout the period 1920–63 the spawning stock biomass of cod was probably less than 150 000 t, now considered by ACFM the 'lowest desirable level' but was higher for haddock, of the order of 150 000 t compared with the 'lowest desirable level' of 100 000 t.

The irony of this story is that throughout the whole period of evolution of the conservation policy ACFM was recommending that the high rates of fishing mortality on these two stocks should be reduced but no action was taken, either then or subsequent to the adoption of the policy. As already noted, the TACs adopted corresponded to maintaining or even increasing rates of fishing mortality. Action was taken when ACFM expressed its concern about the low level of the spawning stock biomasses. Ministers probably recalled the collapse of the stock of North Sea herring and did not wish to be held culpable for the collapse of these two stocks. Unfortunately, the manner in which ACFM has given its advice has led to the perception by ministers, administrators and the fishing industry that if the measures taken are successful in rebuilding the spawning stock biomasses, recruitment will increase and future catches be of the same magnitude as those taken in the period 1964–86. As there is no relationship between the size of the spawning stock biomass and the number of recruits for these two stocks, this is a total misconception.

Even if fishing mortality rates had been reduced, as advised earlier, the same sequence of events would have occurred because if large numbers of recruits do not enter the fishery, large numbers of fish cannot be caught. The only difference would have been that the decline would have happened over a longer timescale because there would have been more older fish to provide a buffer; in 1988–89, 64% by number of the catches of North Sea cod and haddock consisted of fish less than three years old, which means that catches closely reflect recruitment.

Success or failure?

The two best indices of the success of failure of the conservation policy from a practical point of view are total catches and catches per unit effort. The summarized data (Table 7.23) show that, for the stocks analysed, there have been increases in TACs and catches between 1982 and 1992 (1990 for catches) for more stocks than there have been decreases. In contrast, catch per unit effort has fallen for nine stocks but increased for only five. It could therefore be claimed that the conservation policy has been moderately successful and in one respect, the recovery of the North Sea stock of herring, very successful, although stability has not been achieved. The often repeated statements that the CFP is a policy in crisis and that it has been a total failure are not supported by the facts. For many sectors, such as those fishing for herring, mackerel, plaice and sole, with the notable exception of the stock of western English Channel sole, the policy has been ostensibly very successful. It is only for the sector fishing the North Sea and west of Scotland stocks of cod and haddock and, to a lesser extent, North Sea saithe, that the policy has apparently been a total failure. Examination of the facts shows that this was not a failure of the policy. But because the fisheries on these stocks are so important to the British fishing industry, these failures have led it to describe the whole of the CFP as a total failure. An unemotional evaluation of the policy, based upon facts, shows that it is not.

The conservation policy may not have been the unmitigated disaster that most fishermen in the Community claim it to be but could it have achieved more? In 1983 the two main characteristics of the fisheries of the Community were the capture of too many small fish and the high rate of catching of all sizes of fish. For the reasons explained in detail in Chapter 8, reducing both of these would result in higher catches and catches per unit effort in the long

Table 7.23 Summary of increases (+) and decreases (−) in TACs, EC catches and catch per unit effort (CPUE) for stocks for which analytical assessments are available.

Stock	TACs	Catches	CPUE
Herring, IV, VIId	+	+	+
Herring, VIa (North)	−	−	+
Herring, VIa (South)	+	+	?
Herring, Clyde	0	−	?
Herring, VIIa	+	+	?
Herring, VIIg,h,j,k	+	+	?
Mackerel, western	+	+	−
Cod, IV	−	−	−
Cod, VIa	−	0	−
Haddock, IV	−	−	−
Haddock, VIa	−	−	−
Saithe, IV	−	−	−
Saithe, VI	+	+	0
Whiting, IV	−	+	+
Plaice, IV	+	+	0
Plaice, VIIa	0	0	+
Sole, IV	+	+	−
Sole, VIIa	+	+	+
Sole, VIId	+	+	−
Sole, VIIe	−	−	−
Number of increases	10	11	5
Number of decreases	8	7	9
Number of no changes	2	2	2
Total	20	20	16

0 = no change; ? = data not available.

term, as well as greater stability. Have there been improvements in these two respects?

The main means chosen by the Community to reduce catches of small fish has been to increase minimum mesh sizes. As described in Chapters 5 and 6, although minimum mesh sizes have been increased in some fisheries, the amounts by which they have been increased have been minimal, it has taken inordinately long to make the changes and any potential benefits have been negated by both the difficulty of enforcing the regulations and changes in gear design. This part of the policy has been an almost total failure.

Concerning fishing mortality, for the majority of stocks rates have increased (Table 7.24). Based on this criterion the policy has also been an almost total failure. As described in Chapter 6, the reason why rates of fishing mortality have not decreased is that the Council has failed to adopt TACs in accordance with the scientific advice. For other stocks catches have consistently

Table 7.24 Comparison of the annual rates of fishing mortality, expressed as annual percentages, in 1983 and 1990 on those stocks in Community waters for which data were available.

Stock	1983	1990	% change
Herring IV, VIId	27	28	+ 3
Herring VIa (North)	38	22	− 42
Herring Clyde	25	15	− 40
Herring VIIa	16	20	+ 25
Mackerel (Western stock)	19	24	+ 26
Cod IV	59	55	− 7
Cod VIa	55	56	+ 2
Cod VIIa	57	61	+ 7
Cod VIIf,g	49	65	+ 33
Haddock IV	61	61	0
Haddock VIa	39	53	+ 36
Whiting IV	51	54	+ 6
Whiting VIa	41	51	+ 24
Whiting VIIa	66	64	− 3
Whiting VIId	59	69	+ 17
Whiting VIIf,g	82	78	− 5
Saithe IIIa, IV	46	47	+ 2
Saithe VI	27	42	+ 56
Plaice IV	36	43	+ 19
Plaice VIId	39 (1984)	40	+ 3
Plaice VIIa	42	46	+ 10
Plaice VIIe	46 (1984)	34	− 26
Plaice VIIf,g	39	43	+ 10
Sole IV	38	42	+ 11
Sole VIIa	29	34	+ 17
Sole VIIf,g	30	36	+ 20
Sole VIId	37	46	+ 24
Sole VIIe	36	38	+ 6
Sole VIIIa,b	29	41	+ 41

Source: Reports of ACFM for 1990 and 1991.

exceeded the TACs; examples are western mackerel (Table 7.3), North Sea sole (Table 7.14) and both stocks of English Channel sole (Tables 7.16 and 7.17). Both factors have resulted in rates of fishing mortality remaining, at best, almost unchanged, but in most cases, increasing substantially. Only for the stocks of herring VIa (North) (west of Scotland) and Clyde and for plaice VIIe (western English Channel) have rates of fishing decreased significantly (Table 7.24). In this respect also the conservation policy has been almost a total failure.

If the means by which increases in catches and catches per unit effort could have been achieved have not been implemented, any increases in catches cannot be claimed as demonstrating the success of the conservation policy. The converse, of course, is that any decreases in catches cannot be attributed to its failure. So why have TACs and catches varied from year to year? At the present very high rates of fishing on most Community stocks, catches mainly reflect the number of young fish entering the fishery each year, the recruitment. If recruitment rises, catches increase, which has been the general situation for the stocks of plaice and sole. If recruitment fluctuates about some long-term average, catches fluctuate correspondingly. If recruitment declines, as it has done for North Sea cod and haddock, catches fall.

Catch per unit effort fluctuates for the same reason but is more likely to decline because it is also affected by rates of fishing mortality, which, as already shown, have risen for most stocks.

The popular perception that the conservation policy has been an abysmal failure rests essentially on what has happened to two stocks, those of North Sea cod and haddock, although the west of Scotland cod and haddock stocks exhibit much the same features, for the same reason. The data for all other stocks show that the policy has been apparently successful in maintaining or increasing catches. However, if implementing the conservation policy is taken as putting into effect the scientific advice, one of the essential elements of that advice, to reduce rates of fishing mortality, has scarcely been implemented at all. The other essential element, to reduce catches of young fish by increasing minimum mesh sizes, has been implemented at the slowest possible pace (Chapter 6). It could be said with some justification that the Community has never attempted to implement its conservation policy and that the majority of the regulations adopted have been little more than window-dressing. The overriding factor in determining the regulations has been socio-economic considerations, maintaining catches in order to maintain incomes. Because this ignores the fundamental principles of fish stock management, it has inevitably led to problems but the blame for this must lay with those who argue for high TACs and delays in implementing larger minimum mesh sizes, the fishing industry.

Control and enforcement

Political considerations

The Council adopted the first control regulation, Regulation No. 2057/82, before the completion of the CFP on 25 January 1983. As concluded in Chapter 3, the reason for which the Council was able to agree was because the regulation gave no effective powers to the Commission. The political objective was to establish a system of control and enforcement without conceding any competence to the Commission. That this meant that the system would be largely, if not totally ineffective, was almost certainly the objective of most Member States. The major differences between the proposal of the Commission for the replacement to Regulation No. 170/83 and Regulation No. 3760/92 confirm this conclusion. Article 2, paragraph 2, of the proposal laid down the main objectives of a Community monitoring scheme. They were to:

- 'define the most appropriate levels for inspection tasks and ensure coordination thereof;
- determine the means available to persons appointed by the Commission to ensure the effectiveness and transparency of inspections by national departments, in particular guaranteeing the right to intervene without prior notice;
- provide for the obligation of Member States to incorporate into their legislation a system of graded and deterrent penalties over and above the forfeit of the economic gain derived from the infringement;
- permit the use of new technical resources'.

Compare this with the wording of the relevant article, Article 12, adopted:

- 'To ensure compliance with this Regulation, the Council, acting in accordance with the procedure laid down in Article 43 of the Treaty, shall install a Community control system which shall apply to the entire sector.'

The difference in emphasis is further heightened by the fact that the article in the proposal appeared under a separate title, Title II, 'Management and Monitoring of Fishing Activity' whereas in Regulation No. 3760/92 the article on monitoring appears in Title III, 'General Provisions'. A title represents a major section of a regulation, indicating its importance, so no better comment could be made upon how little the Council regards effective control and enforcement.

The Commission has made a proposal for a new control regulation. This is based upon the provisions of the revision of the basic regulation proposed by the Commission and was on the table in December 1992. It was not adopted.

The major political objective has obviously been to avoid transferring any effective powers for control, monitoring and enforcement from the Member States to the Commission. This part of the policy has been only too successful. Ironically, this vital defect can now be presented as a major virtue, that of upholding the principle of 'subsidiarity'.

Practical considerations

Effective control and enforcement depends upon:

- comprehensive but clear, simple to enforce legislation;
- strict enforcement, leading to a high probability of detection of infringements:
- penalties which are a deterrent.

To what extent have these three objectives been met by the Community system?

Legislation
The relevant Community legislation falls into two categories, that laying down what is to be enforced, the regulations concerning TACs, conditions of fishing

and technical conservation measures, and those describing how it is to be enforced, the control regulations.

As already described in Chapter 5, many of the conservation regulations are complex and therefore difficult to enforce. It is permitted to use nets having different minimum mesh sizes, the use of each being associated with minimum percentages of target species and maximum levels of protected species but as there is no restriction on the number of nets of different mesh sizes which can be carried, it is difficult to link catches with the appropriate minimum mesh size. The majority of seasonal restrictions on fishing in certain areas for specified species provide derogations which permit fishing for other species, making inspection at sea rather than aerial survey essential for enforcement. Even in those cases in which the legislation is specific and unambiguous, such as percentage by-catch limits, determining those limits requires exhaustive sampling of the catches. This limits the number of vessels which can be examined. Also, prosecutions are made only when the percentage limit has been so greatly exceeded that the enforcement authorities consider that a prosecution has a chance of succeeding.

In contrast, the control regulations are clear and specific as to what is required of the authorities of the Member States, in particular concerning the recording of catches of species subject to TACs. Regulation No. 2241/87 provides for the completion of logbooks, the recording of landings and the submission of the data to the Commission in order that the uptake of quotas can be monitored. As loopholes have been detected, such as problems concerning trans-shipment and delays by a Member State in reporting landings in its ports by vessels registered in another Member State, they have been rapidly closed. There is also specific legislation on such matters as the marking of vessels with their registration numbers, the measuring of mesh sizes and of by-catches.

But this is all paperwork, whose accuracy is very difficult to check and therefore easy to falsify. It provides more the appearance of control and monitoring than the reality.

Thus, although the procedures of enforcement are precisely described it is frequently at best, very difficult, often impossible to ensure that the legislation is respected.

Detection

The vastness of the problem

Monitoring and controlling the activities of fishing vessels is a vast problem because the area of sea involved is enormous and the number of vessels and ports at which landings can be made very large.

The Community fleet, excluding Greece and Italy and making allowance for Spanish vessels based in the Mediterranean, consists of approximately 50 000 vessels. Assuming that each vessel makes one fishing trip a week for 50 weeks a year, there are some 2.5 million landings each year to monitor. Although the majority of landings are concentrated in a relatively small number of ports, there are thousands of ports at which landings can take place; there is no Community legislation which restricts when or where vessels may land,

Table 7.25 Summary of resources devoted to control and enforcement by Member States in 1990.

Member State	Full-time land-based staff	Vessels (V)	Days at sea (D)	D/V	Aircraft
Belgium	1	4	48	12	0
Denmark	145	7	466	67	0
France	(a)	8	1041	130	(b)+
Germany	47	11	1065	97	0
Ireland	7	5	921	184	+
Netherlands	180	12	1346	112	0
Portugal	12	26	2365	91	(c)2
Spain	17	25	2528	101	0
United Kingdom	180	21	3190	152	(d)3

Notes:
(a) Control and enforcement is the responsibility of the Secretariat of State for the Sea whose staff carry out a wide range of activities connected with maritime affairs.
(b) Approximately 5 000 hours a year.
(c) Approximately 200 hours a year.
(d) Dedicated to control and enforcement.
+ Numbers not specified.

Source: Report on Monitoring Implementation of the Common Fisheries Policy: Commission document SEC (92) 394 final.

although the Netherlands has national legislation on this issue. This situation facilitates the infringement of legislation. As part of its first proposal on conservation measures, the Commission proposed that landings should be restricted to specified ports but this was rejected by the Council.

The fishery zone of the Community is vast. Excluding the Mediterranean Sea, the area of the fishery zones of Member States is approximately one million square kilometres. Based on the figure of 50 000 vessels, this means an average of one vessel for every 20 square kilometres of sea. However, there are three factors which simplify the task of enforcement. All vessels are not at sea at the same time, much of the area is too deep to be fished and the pattern of fishing is not random, in particular many vessels fishing close to the coast. Nevertheless, once a vessel registered in a Member State has sailed it is under no obligation to report its position to any enforcement authorities while it is fishing in Community waters; before it can be controlled, it has to be 'found'.

The lack of resources

Not surprisingly, the resources devoted to control and enforcement vary considerably between Member States (Table 7.25). However, these statistics conceal the effectiveness both of the resources themselves and the manner in which they are used. Many of the patrol vessels are not suitable for the task and they are often not available, either because the budget is insufficient to keep them at sea or because other duties have priority over fisheries enforcement, which is reflected in the average number of days at sea for each vessel.

The Commission has a staff of 19 inspectors who, under paragraph 3 of Article 12 of Regulation No. 2241/87 are '. . . entitled to be present, as deemed necessary by the Commission, at inspections and monitoring carried out by national departments'. But they have no legal powers to enforce legislation. Their role has been described as that of 'looking over the shoulder of national inspectors'.

Training and commitment
The chance of detection of an infringement depends not only upon the resources deployed but how effectively the staff are trained and the equipment deployed. In its 'Report on Monitoring Implementation of the Common Fisheries Policy' the Commission notes that in many Member States there is a lack of political commitment to enforcement with the consequence that staff are poorly equipped and trained and insufficiently motivated. This situation reflects the historically different approach to enforcement in the Member States. The system of minimum mesh sizes and minimum landing sizes is what might be described as a 'northern' approach to conservation, based upon management of the fish stocks according to biological principles. In contrast the 'southern' approach has been to regulate fishing activities by stipulating the size of vessels which can be used in certain areas, the species which may be caught and the fishing gear which can be used and the periods in which fishing may take place. The objective of the latter type of legislation is essentially social, to divide available catch possibilities between many fishermen and to limit competition between them. Under the 'southern' regime, the primary responsibility of those who are now responsible for control and enforcement was to facilitate fishing activities within a framework which was readily comprehensible and therefore acceptable to the fishermen to which it applied. Now these officers have to apply a new type of regime whose basis neither they nor the fishermen understand. In Spain, the situation has had to change the most because the fishing industry was virtually self-regulated by the 'Cofradias', the fishermen's organizations. Although this does not excuse the problems experienced in France and Spain, it helps to explain them.

The Community system of enforcement makes no provision for international inspection; patrol vessels and aircraft may operate only in or over their national waters, preventing sharing of facilities and limiting co-ordination of activities. In this respect the situation is worse now than prior to the extension of 200-mile fishery limits when there was an international scheme of enforcement under NEAFC under which any vessel nominated as a NEAFC patrol vessel could carry out fisheries inspections anywhere within the NEAFC region, which was then international waters but are now mainly Community waters. One of the consequences of this situation is that it leads fishermen to claim that enforcement by the authorities of their Member State is much more strict than that by others and also, when fishing outside their national fishery zone, it is only they who are subject to control. This situation results in some sea areas not being patrolled; where there is no agreement on the dividing line between national fishery zones, with two Member States laying claim to the same sea area, termed 'a grey area', instead of both Member States agreeing to patrol the grey area the customary practice is for neither to do so, except in cases of 'hot pursuit'.

As far as the deployment of the Commission's inspectors is concerned, they operate under a system of 'mutually acceptable programmes of inspection'. Many Member States permit the inspectors unrestricted freedom of movement with the possibility to change these programmes as the inspectors see fit but others prohibit any departure from them. As these programmes are often announced to the industry in advance, precautions can be taken to avoid illegal activities during their visits. Furthermore, paragraph 4 of Article 12 of Regulation No. 2241/87 provides a series of tasks relating in particular to security, defence and customs inspections which take priority over fisheries inspections, allowing the programmes to be modified with no notice whatsoever.

The probability of being detected

What is the probability of a vessel being inspected? This can be estimated only for inspections at sea for which the Report on Monitoring Implementation of the CFP of the Commission gives the number of inspections in 1990 which totalled 20 347, excluding the waters of French Guyana. As already estimated, Community vessels make of the order of 2.5 million trips a year so the probability of being inspected at sea is less than one in 120 trips or alternatively, being inspected only once every two years. Efficiently organized inspection will not be random but concentrated on those areas where it is expected that problems will arise but even so the data show that the chance of being inspected is very small. The number of offences detected was 2393; allowing for the fact that probably some vessels were committing more than one offence, 10% of the vessels inspected were guilty of an infringement; on this basis there could be of the order of 250 000 infringements a year, most of which go undetected.

Information on the number of inspections made in port are not available so it is not possible to make similar calculations. The total number of offences discovered in port and reported for 1990 was 2843, little more than the number detected at sea. As it might be expected that it would be easier to detect an offence in port, this suggests that the level of enforcement is no higher in port than it is at sea unless the proportion of vessels committing an offence is much smaller.

Although the data are presented by country of registration of the vessels and not by Member State, it would be expected that for the majority of inspections in port these would be identical. Assuming this to be the case, there is a wide discrepancy between the number of inspections carried out in port by the Member States (Table 7.26), implying a low risk of being inspected in those Member States.

For one of the offences, undersized fish, it is possible to get an estimate of magnitude of the problem for one species. Of the offences detected in port, 411 related to undersized fish. The reports of the ICES Working Group on the assessment of hake stocks provide data on the length composition of hake which show that, in 1989, 71 million undersized hake were landed,

Table 7.26 Number of inspections in 1990 by country in which fishing vessel registered.

Country	Inspections	
	At sea	In port
Belgium	644	72
Denmark	839	391
France	(a) 2768	281
Germany	446	67
Ireland	253	2
Netherlands	2902	778
Portugal	7349	585
Spain	2459	173
United Kingdom	2285	480
Other	384	14

Notes:
(a) Excluding French Guyana.
Source: As Table 7.25.

representing 41% of the estimated total number landed; these were not marginally undersized fish which might have warranted a warning rather than a prosecution but fish measuring as small as 5 cm. All the undersized fish were landed in France, Spain and Portugal which between them reported 139 offences in this category, 113 in France, 22 in Portugal and only four in Spain. It has to be recalled that the offences are reported by country in which the vessels were registered but nearly all hake is landed in the country in which the vessel is registered. While this situation might not be characteristic of the Community as a whole, it supports the conclusion that the probability of an offence being detected is very low.

Penalties

The third essential element of effective legislation is that the penalties imposed must act as a deterrent. The report of the Commission provides a breakdown of the offences by the action taken but the number of actions taken does not correspond with the total number of offences reported; either there are errors in the data or no action was taken. Assuming the former, 37% of cases resulted in official written warnings being given, 22% in administrative penalties being imposed and 41% being brought to court. Making the latter assumption, no action was taken in 16% of cases, and the other percentages were 31%, 18% and 35% respectively. Thus, even if an offence is detected, the probability of it resulting in a financial penalty is low.

Penalties are fixed by national law and cases are tried by national courts. The report of the Commission provides information on the range of fines provided in national legislation and the actual penalties imposed (Table 7.27).

Table 7.27 Range of penalties provided in national legislation and range of fines imposed in 1990.

Member State	Currency	Maximum penalty	Fines imposed
Belgium (a)	ECU	75 644	236 to 1418
	£	52 766	165 to 989
Denmark(b)	ECU	3167	127 to 30% (b)
	£	2209	89 to 30%(b)
France(a)	ECU	716 to 71 642	716 to 71 642
	£	499 to 49 974	499 to 49 974
Germany	ECU	72 963	2432 to 17 025
	£	50 896	1696 to 11 876
Ireland(a)	ECU	130 141	50 755 to 171 136(c)
	£	90 780	35 404 to 119 377(c)
Netherlands(a)	ECU	10 792	4317
	£	7528	3011
Portugal(d)	ECU	1393 to 27 860	1393 to 27 860
	£	972 to 19 434	972 to 19 434
Spain	ECU	31 397 to 78 493	7849 to 78 493
	£	21 901 to 54 753	5475 to 54 753
United Kingdom(e)	ECU	2867 to 71 679	717 to 64 511
	£	2000 to 50 000	500 to 45 000

Notes:
(a) Gear and catches may be seized for the following offences:

- unauthorized fishing (no quota);
- unauthorized fishing (no licence);
- fishing with unauthorized gear;
- keeping undersized fish on board.

(b) Danish legislation provides for fines to be levied as a proportion, up to 30%, of the market value of the landing to which the offence is related.
(c) Includes value of gear and catches confiscated.
(d) As (a) but immediate seizure of illicit gear when convicted of fishing with unauthorized gear.
(e) As (d) except no provision of seizure of gear when fishing without a licence.
Conversion rate ECU/ £ = 0.6976 (as used in source table).
Source: As Table 7.25.

Although the maximum penalties are very high in some Member States and for some offences, the fines imposed are often only small, notably Belgium and Germany. In contrast, the fines imposed in Ireland were very high but this may reflect the fact that they probably relate to fines imposed upon either Spanish vessels or Irish or British vessels based in Spain and crewed by Spaniards, against which there is much public antipathy which is reflected by the judiciary. In general, the lowest fines were imposed for infringements of the provisions concerning the completion of logbooks, obviously considered of minor importance, although having important implications for the effective implementation of the TAC and quota system, particularly if the number of offences is high; only Ireland imposed a penalty greater than ECU 2867 for this offence,

ECU 50 755. The seriousness of the other offences, in ascending importance of gravity as denoted by the maximum fines imposed but excluding Ireland as being atypical, were keeping undersized fish on board, failure to co-operate with fisheries inspectors, fishing without a quota and, by far the most serious, fishing without a licence.

Conclusions

The control and enforcement of the Community is neither efficient nor effective. The legislation concerning conservation measures, particularly that on technical conservation measures, which has to be enforced at sea is very complex and difficult to understand. In consequence, fishermen are unable to understand and will not respect it and enforcement officers will not be able to implement it for the same reason. The resources available for control and enforcement are woefully inadequate in relation to the vastness of the task, a situation exacerbated by the lack of political commitment in many Member States to effective control.

As a result the probability of a fisherman being apprehended for an offence is very small and, even if he is, the probability of a penalty being imposed which acts as a deterrent to offending in future is small. Fishermen reportedly regard fines as a part of the costs of fishing and are alleged to minimize fines by forming clubs to share the fines. These clubs operate on the basis that when a vessel is arrested at sea it must normally be escorted into port and the patrol vessel must remain in port while the case is heard, allowing the other vessels in the club to fish without fear of apprehension.

Overall assessment

On the basis of whether the conservation policy has achieved its political objectives, the conservation policy can only be adjudged a total success. Access has been regulated. The principle of relative stability has been maintained and Hague Preferences and the transfer of resources have been implemented as planned. TACs have been fixed primarily on the basis of socio-economic criteria. In contrast, it has been an almost total practical failure. The possibilities to reduce significantly catches of small fish and the high rates of fishing have been squandered. There are still as many small fish being caught and the rates of fishing are still as high now as they were in 1983. If these conservation measures had been adopted catches could now be larger than they are and catch per unit effort higher. The policy could not, and never can, result in stable catches, but the fluctuations would have been smaller than they are at present. However, the necessary conservation measures would not only have had to be adopted but also implemented. For political reasons the Community decided not to create an effective system of control and enforcement, which it did successfully, thus contributing to the practical failure of the policy. Even if there had been an effective system, there would still have been severe practical problems of implementing it.

The most notable political success has been the maintenance of relative stability. This is considered as the keystone of the policy and the principle that all Member States except Spain and Portugal were determined to maintain unaltered when the policy was reviewed in 1992. But in order to implement this principle the Community had to adopt a system of TACs and quotas. Even though Regulation No. 3760/92 now provides for allocations of total fishing effort it seems improbable, to say the least, that such allocations will ever be fixed. The system of TACs and quotas will remain in place. As operated by the Community, this system is one of the main causes of the failure of the conservation policy in its practical application. It has been and continues to be one of the major ironies of the conservation policy that the principle which is seen as of such overriding importance by the fishing industry lies at the root of the system which they so detest, the system of TACs and quotas.

Chapter 8

The Conservation Policy: Fisheries Conservation or Fisheries Management?

or Biology v. Economics

Misunderstandings and misconceptions

There exists serious ignorance about the basic principles which underlie fisheries conservation. Despite the fact that there is a vast literature on the subject, much of it written with the objective of simplifying what is a very technical subject, there is still much misunderstanding of what can be achieved and how it can be achieved, and also of what cannot be achieved. Two very recent examples of such misunderstandings appear in the report 'Review of the Common Fisheries Policy' by the House of Lords Select Committee on the European Communities. Paragraph 46 of the report, after defining f_{max}, which it does so incorrectly, describes the implications of fishing beyond f_{max} as follows:

'... if fishing effort is increased beyond f_{max}, too many of the spawning stock are landed and the species cannot reproduce fast enough to keep pace with catches.'

If this statement were correct, once a stock was exploited at a rate greater than f_{max}, it would inevitably collapse because recruitment would continuously fall. Because there is no relationship between the number of recruits and the spawning stock biomass over a very wide range of the latter, for the reasons explained later in this chapter, f_{max} bears no relationship whatsoever to recruitment.

Paragraph 47 contains a similar misconception, stating:

'Another effect of a species having a low spawning stock is that the numbers of young fish being hatched each year (known as 'recruits') becomes very unpredictable.'

Again, for reasons explained later, the level of recruitment is unpredictable whatever the size of the spawning stock biomass.

Given that such misunderstandings occur about basic biological facts, it is no wonder that even greater misconceptions exist concerning the overall objective of the conservation policy. Is it a policy for conservation or a policy for fisheries management? Does achieving the former achieve the latter? What is meant by the term 'conservation'? This chapter examines this issue, first briefly describing the biological principles underlying fisheries conservation and the fisheries models on which fish stock management is based. This provides the essential basis for subsequent chapters.

Basic biological factors

Life history

Except for rays and dogfish, all species of fish have a life history divided into three stages, egg, larva and 'adult'. With the exception of herring, the eggs of all the main commercial species of fish are free floating (pelagic) and drift with the currents. From these eggs hatch small larvae, which do not resemble the adult fish at all and which are also pelagic. It is only after about six months that these larvae change into fish having the adult form. Again depending upon species, but typically at about age two to three years in Community waters, the fish mature and start to spawn. All the species of fish found in Community waters, once mature, spawn once a year. The progeny from each year's spawning is called a 'year-class'.

Growth and mortality

The rate at which fish grow is dependent upon the area in which they live, not the species. For example, all cod do not grow at the same rate; the stock of cod living in the Irish Sea grows more rapidly than that living in the North Sea or west of Scotland. In turn, these stocks of cod grow very much faster than those living in arctic waters.

In Community waters, the weight of an individual fish typically doubles each year in the second and third years of its life. Although the growth rate continually slows down, a fish doubles its weight again in the next three to four years and then doubles it again in the next six years or more. Put another way, a one-year-old fish has the capacity to increase its weight by about 16 times if it survives long enough (Fig. 8.1).

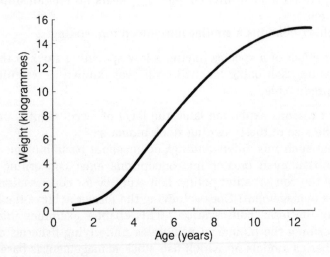

Fig. 8.1 Growth in weight of North Sea cod.

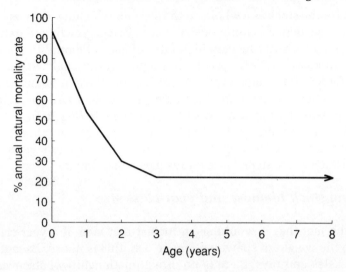

Fig. 8.2 Percentage annual rate of natural mortality of North Sea cod by age.

But many fish do not survive. Even in the absence of fishing, fish are dying all the time, mainly because they are being eaten by other fish but also from disease, about which little is known. Deaths from predation by other fish and marine mammals and disease are termed 'natural mortality'. The rate of natural mortality is highest when the fish are young and most likely to be eaten by other fish. For North Sea cod, the rate of natural mortality is 55% in the first year of life and 30% in the second. Thereafter this rate falls to about 20% a year (Fig. 8.2).

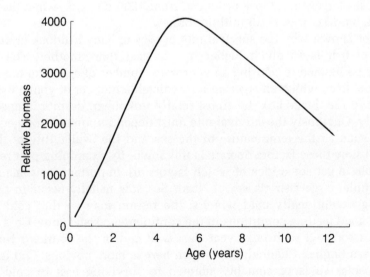

Fig. 8.3 Relative biomass of North Sea cod by age with no fishing.

In terms of the stock as a whole, in the absence of fishing, the weight of a year-class at any age depends upon the balance between growth and natural mortality. Initially, even with the very high rates of natural mortality, the weight of a year-class increases because growth in weight of individual fish more than compensates for losses through deaths. At some point, the two balance and the stock biomass reaches a maximum. Thereafter deaths win and the stock biomass declines. Eventually the year-class becomes no more (Fig. 8.3).

How does fishing affect this situation? Fishing is just another form of death. The higher the rate of fishing the more quickly the size of a year-class diminishes and the earlier stock biomass reaches a maximum.

Spawning stock biomass and year-class size

Despite the fact that it would seem logical that size of a year-class must be related to the weight of spawning female fish, this is not so. Except for sharks, dogfish, skates and rays, there is no direct relationship whatsoever except, of course, if there are no spawning fish there are no juveniles. To illustrate this with an example, in 1970 the stock of mature haddock, the spawning stock biomass, of North Sea haddock was 875 000 t, its highest recorded level. This quantity of mature haddock would have produced about 62 thousand million million (that is 62 followed by 15 noughts) eggs but about six months later there were only 80 thousand million young fish; only one in about 800 000 eggs survived to become a baby haddock. In 1979 the size of the spawning stock was much lower, 103 000 t which would have produced just over seven thousand million million eggs. Six months later there were 67 thousand million baby haddock in the North Sea; approximately one in 100 000 eggs survived. To take another example, the stock of North Sea herring recovered in the early 1980s even though the spawning stock biomass was very small, less than 500 000 t in 1981–83. Yet these spawning stock biomasses produced year-classes larger than those produced from 1989 onwards when the spawning stock biomass was 1–1.5 million tonnes.

It is not known why the survival rate of eggs to baby haddock or any other species of fish is so much higher in one year than another. The process can only be likened to pouring an enormous number of eggs into the top of a 'black box' from which emerges an unpredictable number of young fish. What happens in the 'black box' is almost totally unknown, despite intense scientific study. Obviously the survival rate must depend upon environmental conditions, such as the temperature of the sea and the availability of the right food, but how these factors operate is unknown. By examining past records it is possible to get some idea of which factors are important for some species. For example, large year-classes of North Sea sole usually occur in the years following exceptionally cold winters; the reason may be that cold winters result in good feeding conditions in the continental coastal zone. Cold winters are also associated with large year-classes of cod in the southern North Sea and eastern English Channel. The reason here is more obvious. Cod is a cold-water species so is presumably adapted to surviving best in cold waters. Cod is at the southern extremity of its range in these two areas so survival of eggs and larvae is presumably better when sea temperatures are lower than normal. On the other hand there is no idea why there is an occasional year-class of haddock which is 100 times bigger than the smallest nor why the variability of survival is so much higher for some species, such as cod, haddock and whiting than it is for plaice, for which the variation in year-class size is only three.

Even after the larvae have changed into fish which have the adult shape, the rate of death in the first year and a half still remains very high. Approximately 75% of those which survive the larval phase die in the second six months of their life. In the second year of their life about another 70% die. Nearly all of these deaths are caused by predation by other fish. This has a very important consequence because very small variations in the rate of death from one year to the next make very big differences to the abundance of fish from one year to the next. A large year-class recruiting to the fishery results in a very big increase in catches; conversely if the recruiting year class is very small, catches fall, often devastatingly, as described for the North Sea stocks of cod and haddock in Chapter 7. It required only two to three young to survive each year instead of one, as previously, from the millions of eggs laid to produce the gadoid outburst.

Why, then, do fishery scientists place so much emphasis on maintaining minimum spawning stock biomasses? There are two, possibly three, reasons for this. First, clearly the smaller the size of the spawning stock biomass, the higher the proportion of eggs and young which have to survive in order to produce an average size year-class. Some stocks, such as the North Sea mackerel, the Atlanto-Scandian herring, which used to migrate between Norway and Iceland, and the haddock on Georges Bank, off the east coast of the USA, have all collapsed, the accepted reason being that they were fished so heavily that too few spawners remained to allow stock replacement. Second, ACFM has set itself guidelines to advise 'within safe biological limits'; spawning stock biomass is the only measure which meets this yardstick. Third, a stock collapse is the ultimate disaster. Fishery scientists, particularly those in Europe, have been profoundly influenced in their outlook by the stock collapses which have occurred and their economic consequences. There is a strong feeling of 'never again'. If the worst does happen, fishery scientists want to be able to feel that they gave adequate warnings. The irony of the situation is that if fisheries were managed to maximize economic benefits, as proposed in Chapter 12, stock sizes would increase and spawning stock biomasses would no longer be critical. Managing fisheries on this basis would also correspond with the advice of rates of ACFM that rates of fishing should be decreased.

In practice, nothing can be done to regulate year-class size because the highly variable rates of survival from eggs to young are determined by environmental factors. Catches from a fishery depend upon a fluctuating resource. Man has no control over these fluctuations. For this reason TACs fluctuate so much from one year to the next, a fact which fishermen find difficult to understand or perhaps only pretend to do so as an argument to use against TACs. All that man can do is to manage the resource in order to maximize catches from this fluctuating resource.

Fisheries science

Fisheries models

What needs to be known
Scientific advice on fishery management is based upon mathematical models into which are fed all the factors which determine the catch from a fishery for any species. These factors are the number of young fish entering the fishery (the recruits), the growth rate of the species and the rates of natural mortality and fishing. But it is very important to understand that man cannot manage all of these factors. He cannot manage either growth or natural mortality rates and he cannot manage the number of young fish which enter the fishery each year. The only factor which man can manage is the rate at which he catches fish. This can be done in two ways: (1) by not catching small fish and (2) by regulating the rate of fishing on fish which are large enough to be caught by the fishing gears in use.

Yield per recruit models
The first mathematical models to be developed are called 'yield per recruit models' for the reason that they show only the changes in yield for each recruit entering the fishery resulting from varying the rate of fishing or the age at which fish first start to be caught. When these models were developed, it was not possible to estimate the number of recruits. The growth rates used in these models are based upon average growth rates, described by mathematical formulae, and the age at recruitment is determined from the mesh size and selection factors (see Chapter 9). As their name implies, it is not possible to calculate total catches for a particular year from these models although it is possible to estimate long-term average catches if an estimate of average recruitment is available.

Total allowable catches
The model which is used for calculating total allowable catches uses estimates of the actual number of fish in the sea, including the number of recruits. It is for this reason that it is possible to predict catches using them. The model also uses observed weights of fish at age, not average weights at age based upon formulae, and the observed proportion of fish caught at each age instead of selection factors. They therefore reflect more precisely events in the sea. The same data can be also used to calculate yield per recruit, providing more realistic results than the earlier yield per recruit models.

There are many problems in collecting the data necessary to calculate TACs, which are examined in detail in Chapter 9.

Multispecies models
The models described so far are termed 'single species models' because they assume that each species in each area lives independently of all other species in the same area and does not interact with other species. This is obviously nonsense; some species of fish are the prey of others which are in turn the prey

of bigger species. As who is predator and who is prey is often determined by size, a species which is prey when it is small may become a predator when it is large, even eating the species which preyed upon it when it was small. This occurs within the same species; cod, for example, are cannibalistic. This results in a complex food web which, if it is to be modelled, has to be quantified; not only does 'who eats whom?' have to be determined but also 'how much is eaten?'. Because the quantities of prey which are eaten each year by the predators changes with the availability of food, information has to be collected over many years to ensure that it is realistic. In addition to food consumption, the same information needed to calculate TACs is also required for all the species in the model. Multispecies models therefore require the collection of enormous amounts of detailed information.

Because of the resources needed to collect the necessary data and to analyse them, only one multispecies model exists for Community waters, that for the North Sea which includes only nine species. Another exists for the Baltic Sea, where there are only three main species.

The differences between the results from the two types of model are negligible when small changes to the existing regime are evaluated. It is only when the consequences of major changes, such as large reductions in rates of fishing mortality, are examined that big differences between results are found. It has always been admitted by fishery scientists that predictions from the single species models are unrealistic under these conditions, although this has never been a practical problem because administrators have never made large changes. Multispecies models are inherently more realistic because they attempt to take into account what actually happens in the sea. This presents not only the practical problem of collecting all the necessary data, as already described, but much more importantly, a problem of acceptability. Multispecies models are complex. Although the calculation of a single species TAC appears complicated, it can be carried out by hand using a simple calculator, which is how TACs were first calculated. The method is transparent and the results obtained are those expected. In contrast, multispecies models can be run only on a computer. But, critically for the acceptance of the results, these are often very different from those intuitively expected because the species do not interact in an obvious manner. Because the models have to be run on a computer, it is impossible, except for those who wrote the computer programmes, to understand the reasons. An alternative explanation is that the models have been incorrectly constructed. Whatever the reason, results which cannot be readily explained do not make a readily acceptable basis for management. It also makes the models open to easy criticism by fishermen who find management action based on the results unacceptable. In these circumstances it is very probable that it will be difficult to agree to take action based upon the results from these models.

However, the Community has taken some steps to implementing multispecies management. The results from the North Sea multispecies model predict that reducing the abundance of whiting would result in larger catches of other human consumption species, because whiting is a major predator on these. As described in Chapter 5, a derogation to the standard minimum mesh size in the

North Sea and west of Scotland was made in 1992 in order to allow directed fishing for whiting. Also, the by-catch provisions for the Norway pout fishery in Regulation No. 3094/86 were amended by the TAC and quota regulation for 1993 to permit a higher by-catch of whiting during 1983. Scottish fishermen have used the criticisms which can be levelled against the model to oppose these modifications to the regulations. Not only is the fishery for whiting very important for them but the application of relative stability means that, if the results from the model are borne out in practice, a stock in which the UK has a very large share will decrease, to be replaced by stocks in which the UK has smaller shares. There is an irony in this situation because the same fishermen, as well as every environmentalist, use multispecies-based arguments to oppose industrial fishing, asserting that it is self-evident that catching large quantities of sandeel and Norway pout must have a detrimental effect on the fish, birds and marine mammals which feed on these species because their food is being removed.

Managing fish stocks

The effects of fishing

Taking the simplest case in which the rate of fishing is the same on all sizes of fish, when the rate of fishing is low a large proportion of the fish survive from one year to the next. Catches consist of large fish with a high average age but because only a few fish are caught total catches are small. If the rate of

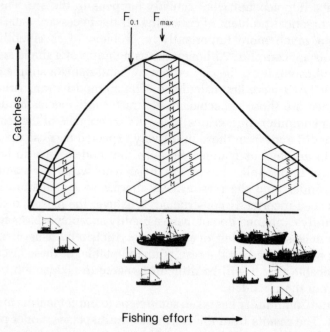

Fig. 8.4 Relationship between catches (yield per recruit) and fishing effort (rate of fishing mortality), F_{max} and $F_{0.1}$. Boxes show how size composition of catches changes as fishing effort increases. L = large fish; M = medium fish; S = small fish.

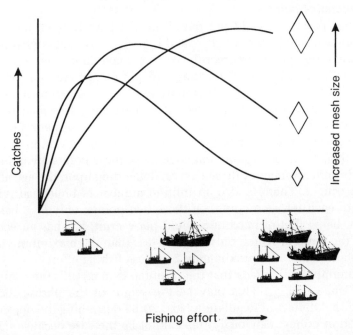

Fig. 8.5 Relationship between catches and fishing effort with increasing mesh size (smallest size at which fish are retained by the fishing gear in use).

fishing is increased the number of fish surviving from one year to the next falls. Consequently, catches consist of a higher proportion of younger, smaller fish than when the rate of fishing was low. But because more fish are caught total catches are greater than when the rate of fishing was low. As the rate of fishing increases this situation continues until a maximum in total catches is reached (Fig. 8.4). As the rate of fishing mortality continues to increase, the number of fish caught continues to increase but because the rate of fishing is so high very few survive to a large size; total catches decrease and consist predominantly of very young, small fish.

The type of curve shown in Fig. 8.4 is called a yield per recruit curve.

The effects of changing mesh size
How does not catching young fish affect this situation? The basic relationship between total catch and the rate of fishing remains unchanged but for any given rate of fishing the total catch is higher the larger the size of fish at which fishing starts (Fig. 8.5). The reason is obvious; the longer that the fish are left in the sea to grow, the more opportunity they have to realize their growth potential and consequently the larger the total catch. There is an upper limit to the size at which fishing must start because otherwise the rate of fishing would have to be so high that it would be impossible to achieve in practice. Not catching small fish is the objective of minimum mesh size regulations but is also achieved by other measures, such as bans on fishing in nursery areas and by-catch limits.

Biological reference points

Two particular points on yield per recruit curves have become termed 'biological reference points'; these are F_{max} and $F_{0.1}$. F_{max} is the rate of fishing mortality at which the yield per recruit is at a maximum (Fig. 8.4). $F_{0.1}$ is technically defined as the point on the yield per recruit curve at which the slope is 10% of that at the origin. It is more easily, if less precisely described as the point at which the yield per recruit is marginally less than the maximum yield per recruit. However, as already noted, the yield per recruit curve changes as the age of the fish on which fishing starts changes or, in more technical terms, the pattern of exploitation changes so these points are not unique. There is an infinite number of these curves, depending upon the age at which fish are first caught. There is also an infinite number of biological reference points, one of each for each curve. Of these two points, only F_{max} has a biological basis, being that of maximum yield per recruit; $F_{0.1}$ has an economic basis, being the point at which only slightly less than the maximum yield per recruit is obtained but at approximately 25% less fishing effort.

It is a commonly held fallacy that the resource is in equilibrium only when it is being fished at F_{max}. This may be the origin of the phrase 'balanced exploitation' in Article 1 of Regulation No. 170/83 describing the objectives of the conservation policy. Obviously this cannot be the case because there are an infinite number of values of F_{max}. More importantly, every point on a yield per recruit curve represents the long-term balance between yield and the corresponding fishing mortality rate.

Regulating fishing effort

The yield per recruit models described in the previous two sections also show that, if the size at which the fish recruit to the fishery is kept unchanged, the yield per recruit can be varied by regulating the rate of fishing mortality (Fig. 8.4). In particular, if a stock is being fished at a rate of fishing mortality greater than F_{max}, the yield per recruit can be increased by decreasing the rate of fishing mortality. This can be achieved in two ways, either directly or indirectly. Direct limitation of fishing effort is implemented by limiting the number of vessels or the number of days which they are permitted to spend at sea or their engine power or the amount of gear which they are permitted to use, or a combination of these factors. The multi-annual guidance programmes are a move in this direction as well as the limitation on days at sea.

Indirect limitation of fishing effort is implemented by limiting catches, the system of TACs. The system works, in theory, as follows. First, it is necessary to be able to estimate the catch which can be taken from a stock for a specified rate of fishing mortality. This catch is then fixed as the TAC. All catches from the stock are recorded. Immediately the TAC has been caught, fishing on the stock is stopped. If all these conditions are met, the rate of fishing mortality corresponding to the TAC will have been achieved.

Fisheries conservation or fisheries management?

Defining fisheries conservation

There exist fundamental differences about what is meant by fisheries conservation and, perhaps more seriously, what fisheries conservation is supposed to achieve. One of the major problems with the term 'conservation' is that the word does not have a unique definition. In its broad, general environmental sense it means 'taking appropriate measures to maintain animals, plants and/or the environment in the state in which they always existed, or to return them to that state'. Thus, there would be little or no dispute in stating that by 'conservation of elephants' is meant trying to restore their populations to the numbers which existed before their large scale slaughter started and then maintaining them at that level. However, this is not what is meant, at least by most people, by the term fisheries conservation, which implies maintaining the stocks while at the same time exploiting them in a rational manner, although this begs the question of what is meant by 'rational'. This situation is changing; there is increasing pressure to consider fisheries conservation in terms of its broader, environmental definition and in the context of the whole marine environment and to question the consequences of their exploitation on other animals such as marine mammals. Importantly for the CFP, this is reflected in Article 1 of Regulation No. 3760/92 which, in describing objectives, refers to '. . . rational and responsible exploitation . . . taking account of its implications for the marine ecosystem . . .'.

The term 'conservation' was not used by the scientists who first investigated the biological principles underlying the management of fish stocks. They were concerned with the efficient exploitation of fish stocks, in particular maximizing the economic returns from fisheries. This is very clearly demonstrated by Graham (1935) who wrote that 'the benefit of efficient exploitation lies more in economy of effort than in increased yield, or preservation of future stocks, though both of these purposes may be served'. Beverton and Holt (1957) wrote a treatise titled *On the Dynamics of Exploited Fish Populations*, which was to become the standard work on the subject for many years and the basis for many conservation measures, without using the term conservation. How then did the term conservation become to be so widely used in fisheries? According to Beverton and Holt it was first used by a series of American researchers to distinguish between regulations having the objective only of increasing catches and those having the objective of improving the economics of the industry. It is in the former sense that the term fisheries conservation has come to be used. The reason for which fisheries conservation has come to have this meaning is therefore a matter of historic accident.

The yield per recruit models show that there are two mechanisms by which the size of catches can be managed, controlling the amount of fishing and the size at which fish are first caught. In the era of unrestricted fishing in what were then mainly international waters it was politically impossible to agree on limiting the amount of fishing. All efforts were concentrated upon controlling the size at which fish were first caught, essentially by means of fixing minimum mesh sizes. The advice on which these were based was biological and the underlying objective was also biological, to obtain the maximum sustainable yield per recruit. Until the end of the 1950s it was confidently expected that, by increasing minimum mesh sizes as the amount of fishing increased, this objective could be achieved.

For the same historic reasons it is considered by many that technical conservation measures are the only means by which fisheries conservation is to be achieved, regulation of fishing effort not being a fisheries conservation but an economic measure.

What fisheries conservation can and cannot achieve

Fisheries conservation, in the sense defined above, can achieve management of the fish stocks and, assuming effective implementation of the technical conservation measures, result in bigger total catches as shown by the yield per recruit curves. But does this result in profitable fisheries? Fishermen catch fish to make money, the essential requirement of every vessel owner being to make a profit if he is to remain in business. Each vessel owner is interested not in the total catch from a fishery but his share of that catch. The most important measure is his catch rate or catch per unit effort because this represents his profit, the difference between how much it costs to take his catch and the amount for which he can sell it. Taking each point on the yield per recruit curve and dividing total catch by the rate of fishing mortality gives catch per

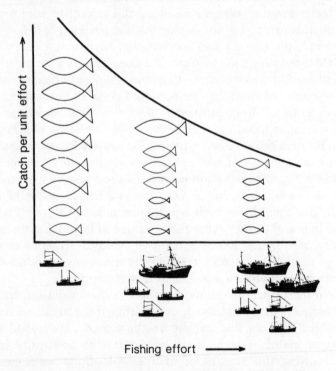

Fig. 8.6 Relationship between catch per unit effort and fishing effort.

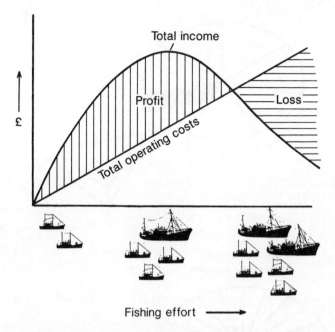

Fig. 8.7 Simplified relationship between total income, total operating costs and fishing effort to show how fishing becomes unprofitable as fishing effort increases.

unit effort (Fig. 8.6). In contrast to the relationship between total catch and the rate of fishing, there is no intermediate rate of fishing at which catch per unit effort is at a maximum; it declines the moment the first fisherman starts fishing. The reason is that once fishing starts there is always less fish in the sea than when there was no fishing so there is less to catch for any quantity of fishing. It has important implications. It means that every additional vessel which enters the fishery lowers the catch rate of all the existing fishermen. To compensate for this, existing fishermen increase their efficiency, if possible. In the short run this increases their catches but it further depletes the stocks so that, in the long run, catch per unit effort decreases. This process is not immediately obvious because the decrease in catch per unit effort following an increase in fishing effort takes perhaps two or three years to happen. It may also be temporarily masked by other events such as the entry of large year-classes into the fishery. The process is insidious but it ineluctably drives fishing effort ever upwards. Every vessel owner has to keep running hard even to remain still.

Another way in which to examine this problem is illustrated simply in Fig. 8.7 in which total catch is equated with total income and a line drawn to represent the increasing costs of fishing. While the line representing costs is lower than that representing income, the fishery is profitable. Where they cross is the point at which income equals costs. In theory no more fishermen should enter the fishery at this point because there is no profit to be made. But they continue to do so because of the time-lag between entry into the fishery and the effects of increased fishing effort becoming apparent. There is an

(a)

(b)   Improved technical
       conservation measures

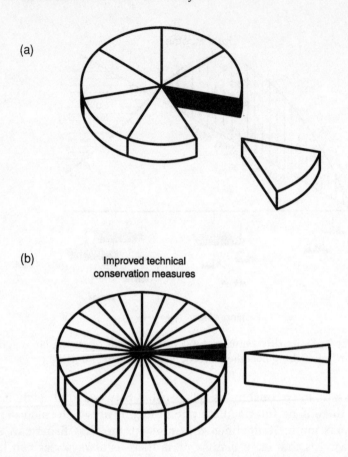

Fig. 8.8 Technical conservation measures may increase total catches ('the cake') but this does not benefit individual fishermen if there are more of them resulting in each getting a smaller 'slice'.

overshoot and the fishery becomes unprofitable. This is made worse by the fact that fishermen who are already fishing do not, and usually cannot, leave the fishery. Instead, they attempt to become more efficient, making the fishery even less profitable.

The consequences of introducing technical conservation measures in this situation are only temporary. If they are effective in increasing total catches the fishery becomes profitable once again. Profits automatically attract new entrants, restarting the cycle until such time as the fishery is once again no longer profitable. So technical conservation measures, if successful, only make the situation worse in the long term and encourage increases in fishing effort. In fact, as the history of technical conservations measures shows, the drive to maintain and increase profitability pushes up fishing effort far more rapidly than it proves possible to introduce technical conservation measures. Put another way, technical conservation measures may increase the size of the cake but if there are more people taking a slice, the size of each person will be no bigger and might even be smaller (Fig. 8.8).

Fisheries conservation can improve the state of the stocks and technical conservation measures have an important role to play in fisheries management but on their own they cannot result in profitable fisheries. For fisheries to be profitable and remain profitable, fishing effort must be limited; unregulated fisheries always become unprofitable.

If limiting fishing effort is essential for achieving profitable fisheries, why has it not been attempted to date in the Community? One of the reasons has already been given, the long tradition of fisheries conservation based upon technical conservation measures reinforced by the biological nature of the scientific advice. But there are other reasons. Limiting the amount of fishing has a direct impact on the activities of fishermen, creating severe political problems, as is being demonstrated now by the measures taken by the Community and national governments, particularly that of the UK. In contrast, although technical conservation measures affect the activities of fishermen, they do not stop them fishing and their effects are neutral between fishermen. The attitude of the fishing industry also militates against change. Most in the fishing industry still believe that its economic problems can be resolved by strictly enforced technical conservation measures. Each fishing vessel owner wants his vessel to be profitable, which is essential if he is to remain in business, but he does not want to be the one who has to cease fishing in order that others can be profitable. Additionally, there is a strong political desire to maintain the communities which depend almost totally on fishing, especially as many of them are situated in regions where there are few, if any, alternative means of employment.

Even when it was decided to regulate the amount of fishing, the means chosen, TACs, had the objective of limiting the rate of fishing mortality on the stocks, not making the fisheries more profitable. Not until the introduction of multi-annual guidance programmes did the Community start to change its attitude but the 'Report 1991 from the Commission to the Council and the European Parliament on the Common Fisheries Policy' does not indicate that the full benefits of economic fisheries management have been grasped by the Commission.

Fisheries management

Fishing is an economic activity. The Community is a capitalist society in which fishermen go to sea to catch fish to earn an income and make a profit. Therefore it is fisheries, the whole complex of fish stocks and the vessels which exploit the fish, which should be managed in order to achieve profits. The appropriate term to use to describe this is 'fisheries management'. Fisheries conservation or, more appropriately fish stock management, is a part, but only a part, of the process in achieving profitable fisheries. That economics is the driving force behind fishing is evidenced, if evidence is required, by the fact that the main argument used by fishermen to oppose conservation measures is the predicted consequent short-term reduction in catches, always equated with reduction in income and their lack of conviction that their incomes will rise in the long term. As described in Chapter 6, socio-economic arguments have been the brake on the conservation policy, the history of the CFP being littered with proposals for conservation measures which have not been adopted, adopted in a modified form or adopted only after long delays because of such opposition. In these arguments it is always conveniently ignored that increases in unit prices of fish as availability falls may often compensate for lower catches by weight but this is part of the price for having no economic input into advice on fisheries management or rather, as the present system should be called, fish stock management.

Fisheries conservation ignores economics; it is not regarded primarily as a means of increasing the profitability of the industry but as a means of improving the state of the stocks, it being implicitly assumed that the former will result from the latter. In consequence, fisheries measures are always considered as remedial, only to be adopted after the stock has deteriorated. In this situation, the measures which have to be taken, such as increasing minimum mesh sizes or reducing the rates of fishing in order to improve the state of the stocks, invariably result in a decrease in catches. The decrease should be short term, to be followed by a long-term increase as a result of implementing the conservation measures. But because the industry is in an economically distressed state before remedial measures are proposed, it cannot sustain the short-term losses and argues for their implementation to be postponed, as has occurred in the case of increasing minimum mesh sizes, or, as in the case of TACs, for increases which will aggravate the situation. Action is often not taken until ACFM advises that the stock is in a critical condition and may collapse. Even then, as demonstrated by the actions taken concerning the stocks of North Sea cod and haddock, the action taken may be far less restrictive than that advised, for economic reasons. Even when the opportunity presents itself to take management measures in order to prevent a stock becoming over-exploited at some time in the future, this attitude persists. For example, the stock of blue whiting is only lightly exploited so NEAFC argues that conservation measures are at present unnecessary. Of course, if and when at some time in the future the stock becomes overexploited, there will be many vested interests opposing conservation measures.

The extension of fishery limits to 200 miles and the establishment of the CFP created the conditions under which the conservation policy could have been developed to achieve its proper economic objectives. That the economic implications of fishery management were so long ignored in the development of the conservation policy has contributed to its failure, a failure exacerbated by the fact that finance was provided under the structural policy for expanding fleets which contributed to the increasing over-exploitation of the fish stocks of the Community, as described in Chapter 2.

The implications of fisheries management

The term 'fisheries management' implies many things. First, and most importantly, it implies that there is a specific objective, or set of objectives which the manager is trying to achieve. It states nothing about what these objectives might be. They could be biological, such as maximum catches, or economic or social or a combination of several. Second, fisheries management implies that there is an overall authority which is capable of taking decisions to achieve the management objectives. Third, fisheries management implies that the management authority has the powers to ensure that the decisions which it has taken are carried out. Fourth, fisheries management implies that the managers control, or attempt to control, all the processes concerned in achieving the objectives as is possible.

The objectives for the CFP are laid down in Article 39 of the Treaty and for the conservation policy specifically in Article 1 of Regulation No. 170/83, now replaced by Regulation No. 3760/90. The objectives described in the Treaty are based upon agricultural practice but, as elaborated in Chapter 11, they can be interpreted in terms of the CFP. However, neither set of objectives is specific and both are internally contradictory, having objectives which are incompatible. In the absence of a decision by the Council as to how they should be interpreted, this renders them useless.

Within the Community there is an authority with power to take decisions, the Council, but as it has not decided on a a set of objectives, its decisions are often incoherent, hindering rather than assisting the rational management of fisheries.

There exist powers to ensure that decisions are implemented. The implementation of Community policies and law is the responsibility of the Commission but Member States have retained responsibility for control and enforcement, which effectively leaves day to day management of the fisheries in their hands. This division of responsibility negates the requirements for effective management.

Thus, although the Community has all the means for managing fisheries it has failed to employ them effectively to date. How fisheries might be effectively managed within the Community framework is examined in Chapters 11 and 12.

Chapter 9

Applying the Science: the Tools of Management, their Application and their Limitations

or Searching for Effective Solutions

The search for effective solutions

The history of the methods used to achieve fisheries conservation is one of continuous improvement to achieve, eventually, the 'holy grail' of fisheries conservation, effective fish stock management. It is also a history of the constant battle between the inventiveness of fisheries scientists to find new methods and that of fishermen to circumvent them. In this chapter the problems associated with the different methods used are described and then possible solutions examined.

Technical conservation measures

Minimum mesh sizes

General problems
The objective of minimum mesh sizes is to increase the yield per recruit by allowing small fish to escape, as explained in Chapter 8 (Fig. 8.5). However, minimum mesh sizes are an imperfect tool for fisheries management. They are not practical for some fishing gears; for example, the mesh of purse seines has to be small enough to prevent fish being meshed, that is, their heads stuck in the meshes, because otherwise the weight of fish would make it impossible to haul the net and might even result in the vessel capsizing. Trawling for pelagic species suffers from the same problem. Implementation of the increase in minimum mesh sizes specified for the fisheries for herring and mackerel had to be delayed because meshing proved a problem (see Chapter 5). In contrast, gillnets are the most selective of all fishing gears because they retain only fish of a certain size, fish below this size swimming through the net and fish above it being too large to become meshed. But gillnets also catch marine mammals and seabirds and there is considerable environmental pressure against their use. For this reason fishing with driftnets longer than 2.5 km is now banned by Community legislation in conformity with the UN resolution prohibiting the use of large, pelagic driftnets. In this context it is worth recalling that the fishery with the longest historic record of sustainable exploitation was that for North Sea herring prosecuted with driftnets and that it was the introduction of trawling and purse seining which brought about its collapse.

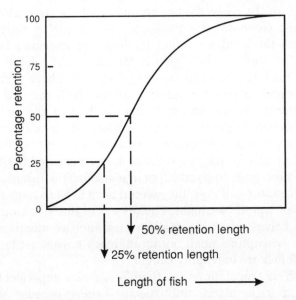

Fig. 9.1 The relationship between the percentage of fish retained and the length of fish for trawls and Danish seines.

Optimization of yields in mixed-species fisheries
The main role of minimum mesh sizes is that in the fisheries using bottom trawls and Danish seines; the following refers specifically to these two gears. As described in Chapter 6, much research was carried out in the period 1946–65 on the relationship between mesh size and the size of fish which could escape, termed 'selectivity'. With increasing size of fish, an increasing proportion of those entering the trawl or Danish seine are retained (Fig. 9.1). This information was essential to developing the yield per recruit models described in Chapter 8 and putting them into practice. Unfortunately for management, the yield per recruit curve is different for each stock of each species because it depends upon the maximum size to which each grows as well as the rates of growth and natural mortality. In consequence, the optimum minimum size of capture is different for each stock of each species. To make matters worse, the selection factor is also different for each. To optimize the exploitation of each species in the same fishing area would require a different minimum mesh size for each species. In fisheries in which a mixture of many species are inevitably caught at the same time, this is impossible. The minimum mesh size has to be a compromise.

Taking for example the fishery for plaice and sole in the North Sea, at the present rate of fishing the 50% retention length which would give the maximum yield per recruit for plaice is 39 cm. As the selection factor for plaice is 2.2, this would mean having a minimum mesh size of 177 mm (39 cm divided by 2.2). As the selection factor for sole is 3.2 the use of such a large mesh size would result in no sole being caught. Consequently, Community legislation permits the use of a minimum mesh size for sole of 80 mm. As sole and plaice occur on the same fishing grounds and are inevitably caught together, plaice are sub-optimally exploited. Much plaice caught must be discarded because the legal minimum landing size provided in Regulation No.3094/86 is 27 cm and is even higher in Danish and Dutch national legislation. The same situation applies in the main demersal fishery in the North Sea and west of Scotland in which a mixture of cod, haddock and whiting are caught. The optimum minimum mesh size for cod at the present rate of exploitation is 290 mm. Using this minimum mesh size would result in catching no haddock or whiting, irrespective of whether the selection factor for haddock is 3.4 or 2.3 or that for whiting is 3.8 or 3.1 as discussed in Chapter 6.

For minimum mesh sizes to be effective in managing the fisheries for species such as monkfish, skates and rays, they would have to be so large that no other species whatsoever would be caught. The reverse of this situation is the problem of exploiting fisheries for very small species, such as prawns, shrimps and *Nephrops* for whose capture small minimum mesh sizes must be used but in which very small fish are inevitably caught.

Minimum mesh size regulations are, therefore, a very imperfect tool for fisheries management in a mixed, multispecies fishery because the standard minimum mesh size has to be a compromise. Consequently it is never possible to maximize the catches of all species in mixed demersal fisheries by using minimum mesh sizes.

The effects of gear construction

Except for investigations on the effect of the length and circumference of lengthening pieces, described in Chapter 6, little research has been done on this subject. The eleventh modification to Regulation No. 3094/86 limited the number of meshes in the circumference of the lengthening piece to a maximum of 100 for Danish seines and trawls having a minimum mesh size of 90 mm or greater. It remains to be seen how effective this measure will prove. It is inconceivable that fishermen were not well aware of the effects of using long lengthening pieces with a large number of meshes in their circumference and presumably it was economic pressures which led them to continue using them. As these circumstances have not changed there still remains a strong incentive to circumvent the latest regulations. Doubtless what was discovered about the effect of lengthening pieces represents the tip of an iceberg. Doubtless also, fishermen will continue to use their ingenuity to frustrate the intention of the regulations by making modifications, which may not even be illegal, if they consider that it is their short-term interests to do so. This also throws the whole issue of minimum mesh sizes as an effective means of fisheries conservation into doubt.

Enforcement

The general problems of enforcement have been described in Chapter 7. The major specific problem concerning the enforcement of minimum mesh size is the difficulty in determining whether a fisherman has respected the percentage species composition which applies to each minimum mesh size, as already described in Chapter 5.

Minimum landing sizes

The objective of minimum landing sizes is to discourage fishermen from using nets of less than the legal minimum mesh size and avoid areas where small fish predominate. Minimum landing sizes are usually, but not invariably fixed in relation to the 50% selection point or, sometimes, the 25% selection point (Fig. 9.1). This means that a mesh of legal minimum size will always let some legal-sized fish escape. There is, therefore, an economic incentive to use nets of less than the minimum mesh size or means by which to diminish the effective mesh size. Similarly, fishermen will fish in areas where small fish are abundant if the catch of legal-sized fish makes it economically attractive to do so. The example of fishing for sole in the continental coastal zone described in Chapter 6 illustrates both these points, except in this case the retention of the derogation of 75 mm made the situation legal.

It is questionable whether legal minimum landing sizes do achieve their objective. All that can probably be argued in their favour is that, if they did not exist, there would be no disincentive whatsoever to use nets having under-sized meshes because they prevent the landing of undersized fish. These fish have to be discarded; few if any survive. The conservation benefits of minimum mesh sizes must, therefore, be very doubtful.

By-catch limits

By-catch limits have the objective of permitting industrial fisheries while limiting the catch of human consumption species. Being based upon a fixed percentage of the total catch, which is never varied, they suffer from the disadvantage that the upper limit to the total by-catch is determined by the catch of the target species. To be effective the percentage limit should be determined by the maximum total catch of the by-catch species required to meet the conservation objective. The classic example of this situation is provided by the situation in the herring and sprat fisheries. The regulations provide for a 10% by-catch of herring in catches of sprat. In the 1970s one of the major impediments to the recovery of the stock of North Sea herring was considered to be catches of juvenile herring in the sprat fisheries. In the period 1976–81 annual catches of sprat averaged 370 000 t so there could have been legal by-catches of up to 37 000 t of juvenile herring. Since 1985 catches of sprat have averaged only 39 000 t so a 50% by-catch now would have the same consequence in terms of the total catch of herring and probably far less serious consequences in terms of the conservation of the stock.

By-catch limits also suffer from the practical disadvantage of having to be determined by sampling. This has two consequences. First, because sampling is very time-consuming, only a limited number of inspections can be carried out. Second, the results from sampling can be easily contested in court because it can always be questioned whether sufficient samples were taken and whether they were truly representative of the total catch. Consequently, prosecutions are brought only when the legal limit is so heavily exceeded that it is thought that the case will be successful.

Seasonally closed areas

These are potentially one of the most effective conservation measures in protecting juvenile fish concentrated into nursery areas but only if there is a total ban on fishing. Otherwise much, if not all, of their effectiveness is lost. If fishing for one species only is banned, fishermen will continue to fish in the area if they consider it economically worthwhile to do so. The species to which the ban applies will be discarded in order to conform with any by-catch provisions in the regulations. These issues have already been considered in Chapters 5 and 6.

Fishermen frequently argue in favour of seasonally closed areas in order to give time for an area to recover from fishing or to protect spawning fish. Neither of these measures is found in Community legislation for the following reasons. Although the abundance of fish might increase in an area which is not fished, once fishing were permitted the fish would be rapidly caught. Any protection would be only temporary. It would have potential benefits only if the area were very large, otherwise fish moving out of the area would be caught by boats which would concentrate on its fringes. To be effective the closure period would have to be very long and the area very large and all fishing in the area banned to avoid cheating. To date, the industry has suggested closing areas of a few tens of square kilometres for only short periods. Although the industry has made the suggestion, the vehement opposition to limiting the number of days at sea suggests that there would be similar hostility to seasonally closed areas of sufficiently large size to have an effect, spurred by exactly the same economic arguments. Also, as described in Chapter 6, the history of legislation on seasonally closed areas does not give confidence that the industry would accept areas in which all fishing was totally banned.

A variant of this proposal is to close spawning areas in order to allow fish to spawn. This is based upon the assumption that the more eggs which are laid the greater the number of young fish which will recruit to the fishery, which has already been shown to be false. It also ignores the fact that once a fish is caught, at whatever time of year or at whatever size, it can no longer spawn. However, the argument holds considerable emotional attachment not only for fishermen but also for administrators many of whom are ignorant of the fact that, except for herring, the eggs of most commercial species are pelagic so that once they have been spawned they are safe from any effects from fishing.

The only Community legislation of this type is the seasonal ban on fishing for herring off the coast of north-east England from 15 August to 15 September each year. However, the objective of this provision is not to protect spawning herring but to prevent quotas being grossly exceeded. The concentration of herring in this area at this time permits large catches to be taken in a much shorter time than it takes for them to be reported and the necessary measures taken to stop fishing once quotas have been exhausted. It is interesting to note that the fishermen concerned object strongly to the ban because it prevents them from catching herring of top quality and which would command maximum prices, an economic argument which would apply to any ban on catching spawning or immediately pre-spawning fish.

Indirect limitation of fishing effort

Analytically based total allowable catches

The method of calculating analytical TACs
Analytical TACs are so-called because they are based upon a series of mathematical calculations which comprise the following steps:

(1) determining the number of fish in each age group in the sea at the start of the year to which the TAC will apply;
(2) multiplying the number in each age group by the rate of fishing mortality on that group;
(3) multiplying the results obtained by step 2 by the average weight of each age group;
(4) adding the weights obtained by step 3 to get the total catch.

In practice, ACFM does these calculations for many different rates of fishing mortality to obtain a series of catch possibilities (step 2). One of these is fixed by the Council as the TAC. ACFM does not determine the TACs.

The following provides a much simplified description of how step (1) is done and is illustrated by Table 9.1. Although simplified it contains the essential details for understanding the problems of using TACs as a means of fish stock management. For the older age groups, the total number of fish in the sea at the start of the year to which the TAC applies is derived from the latest catch data available. This information always dates from two years prior to the year to which the TAC applies. The reason for this interval is that it takes one year to collate and analyse the data. For example, the TACs for 1991 were based upon catch data collected in 1989. For each age group, the number of fish caught is divided by the rate of fishing mortality to get the total number of fish in the sea. Taking the TAC for North Sea cod for 1991 as an example, the number of three-year-old cod caught in 1989 was 15 million. The rate of fishing mortality on this age group was 54%. If this were the only cause of death, an estimate of the total numbers of three-year-olds would be given by dividing 15 million by 0.54, which equals nearly 28 million. However, account has to be taken of deaths from natural mortality, which made the figure 31 million three-year-old cod. This was an estimate of the number of three-year-old cod at the start of 1989. These fish formed the 1986 year-class (see Table 9.1) The next step was to estimate how many of these survived to the start of 1991. This was calculated by applying the appropriate rates of fishing and natural mortality for 1989 and 1990. The cod which were three years old in 1989 were five years old at the start of 1991 and it was estimated that only 5.1 million fish survived to this age. This calculation was made for every age group caught in the fishery, from three years old upwards. But because of the unavoidable delay in processing the data and fixing the TACs, this means that, in the case of North Sea cod, this method is applicable only for cod which were five years old or older in 1991.

What happens about younger fish? Their numbers have to be estimated from research vessel surveys. Each year, research vessels from several countries

Table 9.1 The dependence of catch predictions on estimates of recruitment; catch prediction for North Sea cod for 1991. The data on which the catch prediction is based were collected in 1989. The data were analysed in 1990. The estimates of stock size in 1990 and 1991 were calculated from the data for 1989. Numbers based upon estimates of recruitment are underlined. Only fish up to age five years are shown. (See text for detailed explanation.)

Age in years	Year-class (with reference to 1989)	Stock size on 1 January in millions of cod			Year-class (with reference to 1991)
		1989	1990	1991	
1	1988	324.0 (1)	169.0 (1)	293.0 (2)	1990
2	1987	79.0 (1)	126.5	69.3	1989
3	1986	31.0	21.1	47.5	1988
4	1985	16.5	11.0	8.8	1987
5	1984	1.5	6.2	5.1	1986

Notes:
(1) Estimates based upon recruitment surveys.
(2) Average long-term recruitment.
Source: Report of the Roundfish Working Group. ICES C.M. 1991/Assess:4.

participate in, usually, five different surveys throughout the North Sea, fishing not only for baby cod but other species. From these surveys, the numbers of recruits were predicted based upon the relationship between numbers of baby fish caught in previous surveys and later assessments of their actual numbers. Thus, in 1989 it was estimated that there were 324 million cod which were one year old (the 1988 year-class), of which 47.5 million survived to the age of three years at the start of 1991. Similarly, of the 79 million two-year old cod in 1989, 8.8 million survived to the age of four years on 1 January 1989. Account also had to be taken of the number of one-year-old cod which formed part of the catches in 1991. The only way in which to do this was to take the figure for average recruitment of one-year-old cod (293 million) because there were no data from surveys. Table 9.1 illustrates the manner in which the information is compiled.

Having obtained estimates of the numbers of each age group of cod in the North Sea at the start of 1991 the total catches for different rates of fishing mortality were calculated to give the table of catch options presented by ACFM. From this table was selected the TAC fixed for 1991, which had the objective of limiting the rate of fishing mortality on the stock of North Sea cod to the corresponding rate of fishing mortality, 50%.

The problems in collecting the data
There are many problems in collecting the necessary data and making the calculations. Several requirements have to be met. First, for each stock for which an assessment is made, the total catches landings–discards and by-catches–need to be accurately recorded. Second, all of the catches from each different type of fishing method must be sampled in order that the age composition of the total catches can be calculated. Catches have to be sampled to

determine their length composition and samples of material taken from which the age of the fish can be determined. This sampling must be done in every country which fishes the stock. It must also be continuous throughout the year and from year to year because the method depends upon having long time-series of data. Third, an estimate of the rate of fishing on the stock for the final year in which sampling takes place is required in order to calculate stock numbers from catch numbers. Fourth, estimates of recruitment must be obtained from surveys carried out by research vessel surveys. All this is a tall order, demanding enormous, continuous scientific commitment even though it is internationally shared. Not one of these requirements is met for any stock for which the Community fixes analytical TACs. All of the deficiencies are important, some less so than others.

If total catches are not correctly recorded, both stock size and catch possibilities will be underestimated. It is known that there is much misreporting of catches and data on discards are not available except for North Sea haddock and whiting. Stock size and catch possibilities will also be incorrectly estimated if the rates of fishing mortality are inaccurate, being overestimated if fishing mortality is underestimated and *vice versa*. It is particularly difficult to estimate these rates because data on fishing effort is poorly collected. But this has important consequences because, as described, the numbers of fish in each age group in the sea is estimated by dividing catch numbers by the rate of fishing. The most spectacular case in which this occurred was not for a Community fish stock but the northern stock of cod in Canadian waters and on the Grand Bank. Canadian scientists estimated that the fishing rate was 20% a year whereas it was 30% a year. When they discovered their mistake the TAC had to be halved.

However, the most important problem for stocks in Community waters is that they are so heavily fished that catches consist mainly of very young fish; for example, about 40% of the catches of North Sea cod and 55% of North Sea haddock consist of one- and two-year-olds. As illustrated by Table 9.1, their numbers are determined from recruitment surveys, using sampling techniques. Any numbers determined by sampling are subject to sampling error. These errors are large because the lack of resources limits the number of samples which can be taken. Additionally, the numbers of these recruits which survive to the start of the year to which the TAC applies has to be estimated. When calculating the TAC, it has to be assumed that there will be no change in the rate of fishing or the pattern of fishing between the year of sampling and the year to which the TAC applies. This is a problem for all age groups but, as described in Chapter 8, the youngest age groups are subject to very high rates of natural mortality which vary considerably from year to year, adding further to the uncertainties. As shown in Table 9.1, for North Sea cod the numbers of fish from age one to four in the catch prediction are all based upon estimates of recruitment. In 1991 these formed 77% of the TAC. Yet the assessment for North Sea cod is one of the best. For many stocks estimates of recruitment are not available and average recruitment has to be used.

Some of these problems can be overcome by using what are termed 'fishery-independent' sources of information, such as echo surveys and egg and larval surveys. These are particularly appropriate for stocks of pelagic species for which estimates of stock size can be obtained relatively rapidly and easily using these methods. For this reason, catch predictions for stocks of pelagic species are almost certainly much better than those for demersal species.

The problems of controlling the uptake of TACs

The objective of analytical TACs is to limit the rate of fishing mortality on the stock indirectly. The TAC which is fixed is based upon a certain rate of fishing mortality, as described. If all fishing on the stock is stopped immediately that the TAC is taken, the corresponding rate of fishing mortality will have been achieved. TACs, as their name implies, refer to total catches. The fundamental problem in managing TACs to achieve this objective is that the uptake of quotas is regulated under Community legislation by monitoring landings. All discards are ignored. Inevitably the target rate of fishing mortality for all stocks is almost always exceeded because total catches frequently exceed the TAC, often consistently; examples are western mackerel (Table 7.3), North Sea sole (Table 7.14) and both stocks of English Channel sole (Tables 7.16 and 7.17). The system by which TACs and quotas are managed encourages discarding. When fishing is stopped because the TAC or quota for one species is exhausted, Community legislation allows fishing in the same area for other species for which the quotas are not exhausted or which are not subject to TAC and quota to continue. The species for which the TAC or quota is exhausted continue to be caught. The legislation prohibits these fish from being landed so they must be discarded.

That total catches exceed the TACs, for whatever reason, has an insidious impact on management by TACs, as practised by the Community. An option which is very commonly chosen by the Council is to fix the TAC which corresponds to the current rate of fishing mortality equal on the stock. This is termed the *status quo* option. However, if the rate of fishing continuously increases because catches consistently exceed TACs, for the reasons described, the *status quo* option does not correspond to a constant rate of fishing mortality but to one that is continuously increasing. A ratchet effect is created, leading to ever higher rates of fishing mortality. This is most clearly illustrated by the North Sea stock of plaice (Table 6.5) for which the TACs recommended by ACFM corresponded to increasingly higher rates of fishing mortality between 1984 and 1991 and by the eastern English Channel stock of sole (Tables 7.17 and 7.24). The effect is insidious because it is not obvious.

Summary

Analytical TACs are best suited to the management of a fishery for a single species which is long-lived and which does not recruit to the fishery until it is several years old. The fisheries of the Community are the total opposite to this situation. Both in their calculation and in their application analytical TACs represent an imperfect tool of fisheries management for Community fish stocks, especially for the fisheries for a mixture of demersal species where they are fatally flawed. As a consequence of all these problems, the result of fixing a TAC having the objective of reducing the rate of fishing mortality by 30%

might well range from no reduction at all to a reduction of as much as 60%, even if total catches were recorded and fishing stopped immediately the TAC had been exhausted, which is not the case.

But while relative stability remains the keystone of the conservation policy it is improbable that the Community will abandon TACs. TACs are the means which the Community chose for controlling rates of fishing mortality. The overriding reason for which TACs were chosen was because they can be allocated by quotas, the only politically acceptable method permitting the principle of relative stability to be implemented. The system is politically acceptable because it is transparent and, therefore, indisputable. A tonne of fish is a tonne of fish in anyone's language.

Precautionary TACs

As described in Chapter 3, precautionary TACs are those based upon methods other than analytical assessments and form the majority; of the 105 TACs fixed for 1992, 63 were precautionary. The primary objective of introducing precautionary TACs was political; they were necessary in order to implement the principle of relative stability. Without such TACs the resources to which they apply could not be allocated. Nonetheless, they also form an essential element of the conservation policy. If TACs were fixed only for those stocks for which analytical assessments were possible, fishermen from any Member State could fish anywhere for species not subject to TAC. They would have had to discard species subject to TAC, but this would have been detrimental to the effective management of those stocks for which TACs were fixed.

There are three reasons for which the TACs for many stocks are precautionary. First, for many small stocks it is not cost effective to collect all the data required to make analytical assessments; almost the same amount of information has to be collected irrespective of the size of the stock. Second, the number of analytical assessments which can be done is limited by the amount of scientific manpower available. Third, to calculate an analytical TAC it is essential to be able to age the species; at present it is not possible to age hake reliably and impossible to age *Nephrops*.

However, this does not mean that precautionary TACs have no scientific basis, other than historic catches. For many stocks, such as those of hake, megrim and monk, it is possible to determine their state of exploitation using yield per recruit models. The results from these analyses show that the stocks of these species are fully or over-exploited, implying that TACs should not be set higher than recent average catches.

Precautionary TACs do not have the same objective as analytical TACs, that of limiting the fishing mortality rate on the stocks to some specific rate, although if efficiently enforced they would prevent rates of fishing on the stocks increasing. Their main conservation objective is to ensure that analytical TACs are not undermined.

The fishing industry argues that their catches should not be restricted by precautionary TACs because they have no scientific basis. The industry contends instead that they should be set at levels which would result in national quotas never becoming limiting. This ignores the scientifically valid basis for many precautionary TACs. Also, if conceded, it would exacerbate the problem of discarding which results from fishing in an area when the TAC for a stock has been exhausted. This argument, if followed to its logical conclusion, would result in not fixing any precautionary TACs. There would be no valid reason for restricting the catches of Member States which do not now have quotas in stocks for which precautionary TACs are fixed if the quotas of other Member States are not limited. At present, the principle of relative stability prevents this argument from being followed to its logical conclusion. This argument has been made ever more forcibly since the decrease in the size of many analytical TACs.

Direct limitation of fishing effort

Direct limitation of fishing effort implies limiting one or more of the factors which permit fishing boats to catch fish. The most important of these factors are the number of vessels themselves, their size and engine power, the amount of fishing gear which they carry and the number of days which they spend at sea. To date, the Community has adopted three of these measures. Limitation on days at sea formed part of the conditions of fishing for 1991 and 1992. The multi-annual guidance programmes are based upon limitations on tonnage and engine power of fishing vessels (see Chapter 2).

Limiting fishing capacity

The scientific basis
The scientific basis for limiting fishing capacity is the yield per recruit models. If the age at which fish are caught cannot be increased, yield per recruit can be increased by reducing the rate of fishing mortality (Fig. 8.4); this does not exclude modifying the two simultaneously.

However, the yield per recruit models are based upon rates of fishing mortality, not fishing effort or fishing capacity. To have a basis for controlling fishing effort which is equivalent to that for calculating analytical TACs, it would be essential to have a mathematical model which would permit the fishing effort of any vessel to be converted into an equivalent rate of fishing mortality. Such models exist but they suffer even greater disadvantages than the models used for calculating TACs. First, they require as much, if not more data. Second, they are even more time-consuming to operate. Third, and most important, they would have to be continuously updated every time a vessel was modified or used a different gear.

There is no model which permits fishing capacity to be equated with rates of fishing mortality for the obvious reason that fishing capacity is a measure of the potential of a vessel to catch fish. How much fish it actually catches depends upon many factors such as, for example, the amount of time it spends fishing and the quantity of fishing gear used.

Despite this lack of an appropriate model, the Commission is basing its targets for reducing fishing capacity under its multi-annual guidance pro-grammes on the target of achieving F_{max} or $F_{0.1}$ for stocks for which F_{max} is not appropriate. This objective was selected by the group of experts which the Commission convened to advise it on this issue. However, the group took as its targets those which exist under the present pattern of exploitation – that is, the rate of fishing mortality on each age group. The group of experts did not consider how these values might be changed by altering the pattern of exploitation. It also did not consider that the changes predicted in yield per recruit might well be totally unrealistic for some of the very large reductions in the rates of fishing mortality required by these models to achieve F_{max}; for example, a reduction of 76% is required for North Sea cod.

The problems of limiting fleet size
As described in Chapter 2, many Member States have not achieved the limited objectives set for the programmes for 1987–91. It is therefore questionable whether they, or any Member State, will achieve the targets set for the period 1993–96 (Table 2.4). Although multi-annual guidance programmes are legally binding, being decisions of the Commission (see Chapter 1), the Commission has been unable to ensure that Member States implement them. Even if the Commission prosecuted those Member States which have failed to respect their programmes in the Court of Justice, where the Commission would inevitably win any case, it would be a hollow victory because the vessels have continued to fish.

A major defect in implementing the multi-annual guidance programmes to date has been the lack of statistics on fleet sizes. Most Member States do not keep records of the number of small vessels. In no Member States are there records of which registered fishing vessels are actually engaged in fishing. It is thought that much of the reductions which have occurred to date have result-ed from inactive vessels being removed from the registers. Only since 1989 was an attempt made to compile a Community fleet register. Council Regulation No. 163/89 required all Member States to submit detailed informa-tion about all fishing vessels on their registers on 1 January 1989 to the Commission and to update this information monthly. However, the two key characteristics, tonnage and engine power will not have to be recorded in accordance with international standards until 18 July 1994, the date on which the relevant international conventions enter into force.

Practical problems
Direct limitation of fishing effort suffers from the same drawback as minimum mesh sizes and TACs, that fishing effort cannot be targeted against individual species in mixed demersal fisheries; yields from all species cannot be maxi-mized at the same time. To be effective, the cutbacks in fishing capacity need to be targeted on the fleets which are fishing the stocks on which reductions of fishing effort are necessary. At present the rate of over-exploitation is such that an overall reduction in all fishing effort is required but as and when the sizes of the fleets are reduced it will be necessary to target the reductions on those fleets which are producing excessive fishing effort. The Commission has

taken a first step in this direction with the multi-annual guidance programmes for 1993–96. However, the problems of relating rates of fishing mortality to fishing effort, let alone fishing capacity, and the difficulties of attempting to describe on a scientific basis which fleets are causing which components of mortality will make this step increasingly difficult to implement, especially as vessels can switch between fisheries.

As already stated, there is no method by which fishing capacity can be directly related to rates of fishing mortality or fishing effort. Fishing capacity is a measure of the size of the fleets, not of how much they fish or what species they catch. For this reason it is difficult to foresee how the Community might attempt to fix total allowable fishing effort and allocate it between Member States, as foreseen by Regulation No. 3760/92.

Summary
Regulating fishing effort by limiting fishing capacity is an imperfect method of managing fisheries. But a perfect method does not exist. Scientifically it can be likened to using precautionary TACs rather than.analytical TACs.

The major advantage of regulating fishing capacity is twofold. First, it is much easier to control and enforce fishing capacity than TACs and quotas. All fishing vessels will be licensed under the provisions of Regulation No. 3760/92 and it is also foreseen in the proposals of the Commission that their activities will be monitored by satellite. Second, despite the lack of a rigorous scientific model, the fewer the number of fishing vessels, the lower the rate of fishing mortality is likely to be. Ultimately, if a vessel does not exist, it cannot catch fish.

Limitation of days at sea

As described in Chapter 6, limitation on days at sea was first introduced in Regulation No. 3926/90 which fixed the TACs for 1991 and the conditions under which they could be fished. Similar provisions were made for 1992. They were not renewed for 1993 because the UK government's Sea Fish (Conservation) Bill 1992 provides for limiting days at sea for all British vessels, wherever they fish. As these are the vessels which were most affected by the Community legislation, the Council did not consider it necessary to make similar provisions. Limitation on days at sea was used by the Community as a temporary measure in what was seen as the desperate state of the stocks of North Sea and west of Scotland cod and haddock in 1991 and 1992.

Limiting days at sea has the same advantages and disadvantages as regulating fishing capacity. As for regulating fishing capacity, there is no scientific model but on the other hand if a reduction of, say, 40% in the rate of fishing mortality is required, reducing the number of days fishing by 40% is likely to achieve this even if fishermen claim that they fish harder while at sea. That UK fishermen fought so hard against the continuation of the eight-day tie-up measure in 1992 certainly suggests that it limited their fishing activities in 1991.

The major disadvantage with limiting days at sea is that it is economically very inefficient. The fixed costs of the fleet are not reduced, only its variable costs, such as fuel and gear.

Discarding

Discarding, the throwing back into the sea of fish immediately after capture, is a major issue in the Community because of the large quantities of fish which are reported as being discarded. Discards are a waste of resources and undermine the management system, particularly that of TACs. The question requires separate discussion because many of the solutions proposed to rectify the shortcomings of technical conservation measures and TACs have as their primary objective minimizing discards. Fishermen have always discarded fish for which there is no market, either because they were species which were not acceptable to consumers or because they were too small. With the introduction of legal minimum landing sizes, prohibiting the landing of undersized fish was the only method of enforcing the legislation. As already described, Community legislation uses the same basis for enforcing TACs with the same consequence, discarding. It has never been possible to get the Council to adopt any measure which will minimize discarding, such as stopping all fishing in an area once a Member State has exhausted its largest quotas. Additionally, fishermen attempt to maximize the value of their catches by retaining only the largest, most valuable fish, discarding the rest, a practice known as 'high grading'.

Discarding has become an issue only with the introduction of TACs but they are not, as many claim, a direct consequence of the system. High rates of discarding occurred for many years in the North Sea fisheries for haddock and whiting long before the adoption of the CFP or any conservation measures. Between 1924 and 1930 there are records showing that 51% by number and 35% by weight of the haddock being caught in some areas fished by Scottish vessels were discarded. More recent information, obtained from surveys funded by the Commission, show that discarding occurs in many other fisheries. For example, total hauls of herring may be discarded in the fishery for herring of which only the roes are used; the roes are destined for the Japanese market, whose very exacting quality specifications have to be met. Discard rates of 10–15% are more usual in other herring fisheries. In the trawl and Danish seine fisheries of Denmark maximum discard rates of 33% for plaice, 29% for cod and 68% for haddock were observed; estimated total weights discarded were not reported. Curiously these surveys showed that fish much larger than the legal minimum landing size were discarded, cod of up to 41 cm compared with a legal minimum landing size of 35 cm, and haddock up to 35 cm compared with the legal minimum landing size of 30 cm. The reports also show that many undersized fish were caught with what presumably were nets having legal minimum mesh sizes, the smallest fish recorded being 20 cm for cod, 15 cm for haddock and 20 cm for plaice. Similar results were obtained in a survey of the south-west of England demersal trawl fishery.

Possible solutions to the problems

Many suggestions have been made to overcome the problems which arise in trying to implement technical conservation measures and TACs. Many of these solutions are concerned with preventing discarding, particularly in the demersal, multispecies fisheries. The advantages and disadvantages of these are examined in the following sections.

Legislation and enforcement

There is a school of thought which argues that many of the problems arising from implementing management measures could be overcome by having more legislation and more effective enforcement. These subjects have been discussed in Chapter 7 where it was concluded that it would be advantageous to the implementation of the conservation policy if the legislation were simpler and there was less, rather than more. It was also concluded that enforcement at sea is a major problem which cannot be done effectively at a reasonable cost, at least not under the present system in the Community.

One of the most favoured options to prevent discarding is simply to ban it, following the example of Norway. But it is very uncertain how Norway applies its legislation, particularly in respect to landings in excess of quota and undersized fish. Norway has the advantage over the Community that it monitors catches at sea. If a large proportion of small fish are caught in an area administrative action can immediately be taken to ban fishing in the area until the situation changes. Even assuming that the Community agreed a budget for the observers required, history does not provide grounds for optimism that the Council would give the Commission the necessary powers. In 1980 it refused to adopt a proposal which would have given the Commission such powers in order to limit by-catches of herring in the sprat fisheries. Almost certainly the Council would insist upon the Management Committee procedure being followed, creating a delay which would negate the objective.

Even if such measures were implemented, there still remains the problem of landings of undersized and over-quota fish. Under the Norwegian system this is dealt with by a combination of administrative penalties and prosecutions depending upon the gravity of the contravention. Given the poor record of the Member States on enforcement, it seems highly probable that similar provisions would be grossly abused. Also, if all fish caught had to be landed, the TACs and quotas would become exhausted that much earlier, providing a strong incentive for fishermen to cheat. This would be counterbalanced to some extent by the fact that the TAC would be larger if account were taken of all catches. A ban on discards can be enforced only at sea and the ingenuity of fishermen plus the low level of enforcement suggests that fishermen would have little difficulty in circumventing such a ban. Norway has certainly not been 100% successful despite its extremely efficient enforcement service.

Technical solutions

Instead of trying to enforce legislation which makes fishermen discard fish, it would be far better if they could be forced to use gear which would not catch them in the first place. It should not even be necessary to encourage fishermen to do this, let alone force them, because not catching fish which have to

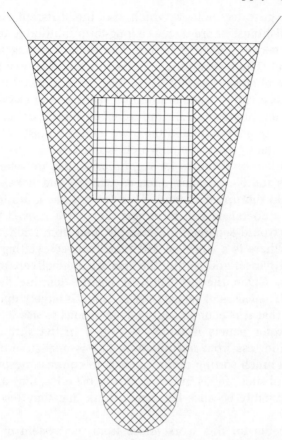

Fig. 9.2 Panel of square mesh netting in a codend of diamond mesh.

be discarded is in their own interest – but the discard data suggest otherwise. The most effective means which has been found to improve selectivity of trawls and reduce catches of undersized fish is to insert a panel of square mesh netting in the codend (Fig. 9.2) or, alternatively, a metal grid, as used in Norway. The netting of such a panel, even less so a metal grid, does not deform under the strains of towing, irrespective of the construction of the gear. Fish can escape undamaged. The British fishing industry has done much work on the use of square mesh panels itself and favours its adoption as a practical solution to the problem. When the Council modified Regulation No. 3094/86 for the twelfth time it made provision for the use of square mesh panels but made their use optional, not mandatory. There appear to be two reasons for this. First, the use of square mesh panels appears to be a practical solution only for fisheries in which roundfish, not flatfish, are the main target species. Second, the Scottish fishing industry argued that if square mesh panels were made obligatory, their minimum mesh size should be 80 mm, otherwise too many saleable fish would escape. As this would have represented a reduction in minimum mesh size, the Commission found it unacceptable and did not propose it.

Developments in gear technology which use the different reactions of different species to fishing gear are seen as a long-term solution to the problem of managing multispecies, mixed fisheries. The Community is funding research on this subject. There exist proven cases in which gear can be adapted to retain one of several species caught, the others being allowed to escape or directed into codends of appropriate minimum mesh size. For example, trawls having separator panels are used in shrimp fisheries; shrimp do not rise when they enter the trawl and are retained in the lower codend; most of the fish rise and escape through the upper codend, which is not fastened. Similar trawls have been designed for the fisheries for *Nephrops* which are retained in the lower codend while the fish species, such as whiting and hake, depending upon area, swim into the upper codend. This codend has a minimum mesh size which permits undersized fish to escape. Even if it proved possible to design gears which would separate, for example, cod from haddock, or haddock from whiting, there is a world of difference between getting successful results on a research vessel and getting the gear used effectively on commercial fishing vessels. Given the problems in legally defining fishing gear, referred to earlier, the success of selective gear depends largely upon whether fishermen consider that it is economically advantageous to use it. The use of trawls having separator panels is widely accepted in the shrimp fisheries because the economic loss from not retaining fish is negligible or even nil; also, their use saves much sorting of the catches. In contrast, despite years of research on the use of such trawls in the *Nephrops* fisheries, they are not used commercially, presumably because fishermen think that they lose too many saleable fish.

Much of the impetus for this work arises from the system of TACs and quotas. The relative proportions of the species caught never correspond with those of the quotas. In consequence, sooner or later a quota of one species is exhausted and discarding has to start, at least under the present system of enforcing quotas. For example, the Commission is funding research on beam trawls with the objective of reducing catches of roundfish because most Member States which have large quotas of plaice and sole, caught by beam trawlers, have small quotas of roundfish. While this makes sense in the context of the present system, it makes economic nonsense to design gear which will let fish escape and which could be retained at no additional cost. For this reason alone it is unlikely that fishermen will readily use such gear if it is successfully developed.

There are two major problems with using gear construction as a means of regulation. The first is that, although a specification for any fishing gear can be written, fishing gears are infinitely variable, every skipper of a fishing vessel probably using his own variant of even a standard design. It is highly improbable that the fishing industry would accept specified gear being laid down by Community legislation with all non-specified gear being prohibited. The second problem is one of enforcement. Not only is the risk of getting caught low, as described in Chapter 7, but the gear would have to be checked in the very confined space of the deck of a fishing vessel – no easy task.

Combined species TACs

By the term 'combined species TACs' is meant any system under which catches of one stock can be debited against the TAC for other stocks in the same area. The term 'multi-species TACs' is often used to describe this system but this leads to confusion between what is meant in this context and TACs based on the results from multispecies models. In the 'Report 1991 from the Commission to the Council and the European Parliament on the Common Fisheries Policy', it was suggested that it should be made possible to transfer species within TACs or quotas for the same area. For example, if a Member State had exhausted its quota for cod it could continue to fish for other species, such as haddock and whiting, debiting any cod caught against the quotas for these other species.

Regulation No. 3760/92 is based upon the ideas in this report but it is not evident whether it makes provision for combined species TACs or not. Explicitly it does not. The regulation does provide for management objectives to '. . . be established on a multispecies basis . . .', which would appear to imply the result of multispecies models. However, the relevant recital states:

'Whereas with a view to effective conservation, the rates of exploitation of certain resources should be limited and may be fixed on an annual or, where appropriate, multiannual basis and/or on a multispecies basis'.

The liaison between multiannual and multispecies-based management, taken in conjunction with the contents of the above-mentioned report, suggest that combined species TACs are meant.

Whether or not the Community now has the legal right to adopt combined species TACs, the implications need considering. A completely unrestricted combined species TAC would result in fishermen landing only the most valuable species. For example, in the mixed demersal fisheries of the North Sea they would land the two most valuable species and discard the least valuable, whiting. Means would have to be found to prevent this. One way would be to limit the quantity which can be transferred and apply weighting factors to discourage fishermen fishing for the most valuable species. For this to be effective, two requirements would need to be met simultaneously. First, the weighting factor applied to excess catches would have to be greater than the inverse proportion of the relative prices per unit value of the species. For example, the ratio of the unit price of cod to whiting in the UK in 1991 was 1:0.41. Each tonne of excess cod would have to be counted at more than 2.44 tonnes (1/0.41). If this were not done, there would be an economic incentive to discard the lowest value species from the start of the TAC year if the expectation were that this could be replaced by catches of higher value species. Second, it would be essential to ensure that those who were responsible for overfishing paid the penalty. This would be possible only in the case where national quotas are allocated to producers' organizations.

A variation of this approach, but one which it has not been suggested that the Community might adopt, is to fix, in addition to the individual TACs for each stock, a combined species TAC which is less than their total. This allows for overfishing of some TACs but at the penalty of a lower total catch. This system was tried by ICNAF which set the combined TAC at 90% of the sum of the individual TACs. However, discarding of the least valuable species would render this system ineffective.

Combined species TACs also ignore both the political and biological reasons for fixing TACs. Politically, permitting transfers would undermine the principle of relative stability. Biologically, it would mean that analytical TACs could not achieve their objective of limiting the rate of fishing on a stock to the level corresponding to the TAC. Both these factors plus the complexity of managing combined species TACs make them unsatisfactory as a means of resolving the problems set by a system of TACs in a multispecies mixed fishery.

Multi-annual, rolling TACs

Regulation No. 3760/92 makes provision for multi-annual TACs. It is assumed that, under these provisions, the Council will adopt legislation which will permit catches which exceed the TAC or quota in one year to be debited against the next year's TAC or quota for the same stock. This would eliminate the need to discard fish for which the TAC or quota is exhausted. Confusingly, this system is also often referred to as 'multispecies TACs'. This suggestion raises three major problems, one of assessment and two of management.

Multi-annual TACs raise serious problems of how they are to be assessed. As already described, at present rates of fishing a high proportion of many TACs consists of fish whose numbers were estimated from recruit surveys

Table 9.2 The problem of fixing multi-annual TACs, illustrating how the number of age groups based upon estimates of recruitment increases with the number of years ahead the TAC is fixed. See also Table 9.1.

	Collect data in 1989	Assess data in 1990	Predict TACs for							
			1991		1992		1993		1994	
Age	*Year-classes*									
0	1989	1990	1991	S	1992	NS	1993	NS	1994	NS
1	1988	1989	1990	AR	1991	S	1992	NS	1993	NS
2	1987	1988	1989	R	1990	AR	1991	S	1992	NS
3	1986	1987	1988	R	1989	R	1990	AR	1991	S
4	1985	1986	1987	R	1988	R	1989	R	1990	AR
5	1984	1985	1986	L	1987	R	1988	R	1989	R
6	1983	1984	1985	L	1986	L	1987	R	1988	R

Key:
For an assessment made in 1990, catch prediction for each year-class in each year based on:
NS: no data (year-class not spawned);
S: no data (year in which year-class spawned);
AR: average recruitment data;
R: data from recruitment surveys;
L: age and length data.

(Table 9.1). A TAC for two years ahead would include fish which had not even been spawned when the assessment was made. The situation becomes progressively worse the further ahead TACs are predicted (Table 9.2). In fact, under these circumstances it would be more appropriate to use the word 'guessed' than 'predicted' because the TAC would have to be based on average recruitment. The assessments would also have to assume no change in the pattern of fishing over the period to which the multi-annual TACs applied. The TACs would also have to be fixed in accordance with a specified, long-term objective. At present, TACs are not fixed on a consistent basis from year to year. Regulation No. 3760/92 does provide for the fixing of long-term objectives but it remains to be seen whether the Council will adopt such objectives and, more importantly, if it does whether it will adhere to them when faced with crises.

Management of multi-annual TACs would require a fundamental change in the method of operation of the Council. At present it is an unwritten principle of the Council that TACs, once adopted, are never reduced during the course of the year to which they apply, notwithstanding scientific advice. This principle is based on the argument that fishermen have planned their activities in accordance with the quotas available and these plans should not be disturbed. Curiously, this argument does not work in reverse. Advice that TACs may be increased is implemented immediately. Even under present circumstances the effect of this principle is to increase rates of fishing mortality, another factor in the 'ratchet effect' described earlier.

There is no reason why this principle should be regarded as sacrosanct but it must raise doubts about how the Council will respond if and when it does adopt multi-annual TACs. The fishing industry will always contend, as evidenced by the situation described for English Channel cod, that it should not be prevented from catching fish which are there. The Council is likely to respond to such pressure and increase multi-annual TACs. There will be no serious danger to the stocks in doing this. The danger arises when the scientific advice is that TACs must be reduced. The situation becomes most critical when recruitment declines over a long period, as for North Sea cod and haddock. If TACs are not reduced in this situation, rates of fishing will increase rapidly with the danger of stocks collapsing.

Multi-annual TACs would be managed by permitting excess catches from one year to be debited against the TAC or quota for the following year. It is possible that a weighting factor might be applied to such catches so that they counted more than their actual weight. This would have the objective of discouraging overfishing and act as a penalty. It would also compensate for the fact that the excess fish caught would have weighed more if they had been caught later. The concept of multi-annual TACs is seductively simple especially when it is presented as 'why should it be illegal to catch fish on 31 December which it will be legal to catch on the following 1 January?'. As for combined species TACs, this ignores the biological objective of TACs, which is to limit rates of fishing mortality in the year to which they apply. The concept of multi-annual TACs is also based upon two assumptions, first that overfishing and underfishing will equalize over the long term and, second, that overfishing will not be excessive. In a situation in which there is excess fishing capacity, the probability that over-fishing will occur is much higher than the TAC will be under-utilized. There would be a high probability that within a few years catches would be debited against the TAC for one or two years ahead. Even though the proportion of the TAC which could be carried forward in order to prevent the situation arising might be limited, experience suggests that once the principle was conceded the percentage would be increased by the Council as it considered necessary in the light of 'socio-economic' circumstances at the time. The argument 'why should it be illegal to catch fish on 31 December which it will be legal to catch on the following 1 January?' would be extended, day by day, *ad infinitum*.

Individual transferable quotas

In a system of ITQs, quotas are allocated to individuals who have the right to buy and sell them. It has been argued that individual transferable quotas resolve all the problems of managing multispecies mixed fisheries. The theory underlying this argument is that, under this system, a fisherman must buy ITQs in order to cover all catches which exceed the ITQs already held and that this would be possible through a market for quotas which would become established. No fisherman would be able to sell his catch without proof that he held the necessary ITQs. It is contended that at least some of those holding ITQs would not fish them but retain them, speculating on the high price which they might expect to obtain towards the end of a year when the demand would become greatest.

The theory ignores the fact that the effective operation of the system is totally dependent upon having a sophisticated on-line quota dealing system providing proof in real time that sufficient ITQs to cover landings were held. It would also require very effective enforcement. Almost certainly it would necessitate limiting the number of ports at which landings could be made and probably the times of landings also. Not only does experience suggest that the Council would not be prepared to adopt such provisions but, for the reasons already described, there is little probability that the Member States could effectively enforce such a system. In the absence of effective enforcement, fishermen would simply land their catches illegally or discard them, especially once the price of the quota exceeded the market value of fish to which it referred. The Netherlands has had a system of ITQs for plaice and sole for many years and has had enormous problems in trying to control catches, problems which still exist despite restricting the ports at which fish may be landed as well as the times.

Additional research

Another approach favoured by the Community is additional research. It is believed that if better, more realistic mathematical models could be developed, fisheries management would be improved. This ignores the fact that the problems which confront fisheries management result not from a lack of knowledge or inadequate models but a failure to apply existing knowledge.

The fundamental principles of fisheries management have been known since 1935; the failure has been to apply them. Developing better models is of no use if the results are not implemented. Economic models, of which there are many, have been totally ignored.

Further research on gear design is also considered as leading to solutions but, as already indicated, this is a solution to a problem which could be resolved more effectively by changing the system of TACs and quotas; also, history shows that fishermen will not adopt the gear unless forced to do so and then will cheat if it is economically advantageous to do so .

Conclusion

Fisheries science has provided a series of models which provide the basis for fish stock management. Their results can be encapsulated in two basic rules, 'don't catch small fish' and 'don't fish too hard'. For historic reasons fisheries management concentrated upon the former rule for several reasons. First, in the era prior to the extension of 200-mile fishery limits it was impossible to obtain agreement upon limiting fishing effort. Second, its implementation had the advantage that it did not discriminate between fishermen; although it meant mainly restrictions on the type of gear which they used, it did not otherwise prevent them from fishing as they pleased. Third, it might also be said that the regulations were difficult to enforce and therefore easy to circumvent, and for this reason it was more easy to reach agreement upon them. However, even if perfect technical conservation measures regulations could be drafted and enforced so that there was not a single infraction, the multispecies nature of the demersal fisheries in Community waters places severe limitations on their efficacy as a means of management.

Control of fishing effort has proved equally ineffective. Indirect control of fishing effort by means of TACs is subject to so many scientific, political and administrative problems that it is highly improbable that they can ever be an effective measure in managing mixed demersal fisheries although they may be much more satisfactory for managing the essentially mono-specific pelagic fisheries. The various proposals to modify their implementation are simply tinkering with a fatally flawed system and ignoring the political realities of decision taking in the Community. These realities are that decisions are postponed until such time as events force them to be taken and that, when they are taken, the legislation negates the achievement of the objectives which the decisions were supposed to attain.

Direct control of fishing effort suffers from the fact that its scientific basis is general, not specific and, like technical conservation measures, it cannot be operated selectively in the demersal fisheries. Of the two means of direct limitation employed by the Community to date, limitation on days at sea has been implemented in a largely ineffective manner and reduction in fleet capacity is only now being seriously attempted. The lack of success to date on reducing fishing capacity by a small amount does not augur well for achieving the much larger reductions required by the multi-annual programmes for 1993–96.

Many of the problems in applying the science stem from the fact that the conservation policy of the Community is based upon a system of fish stock conservation, not fisheries management, Because fish stock conservation cannot result in fisheries becoming profitable there is always economic pressure on fishermen to circumvent legislation simply to earn a living. Technical conservation measures are always going to be a weak instrument in this situation. Concerning TACs, most of the solutions suggested ignore their scientific inadequacies, inadequacies which are exacerbated by the means by which they are implemented under the conservation policy of the Community. They also ignore the objective for which TACs are set, to limit the rates of fishing mortality on fish stocks. All the solutions which involve the transfer of TACs between species and between years ignore the experience that the Council has never adopted effective measures to limit the exploitation of the stocks. Setting an inflexible TAC which applies to a fixed period has the advantage that it is legally binding and all fishermen know precisely what is the catch limit for that year. Once that limit can be varied, the floodgates will be opened and history shows that the Council, under pressure from the industry, will exploit the situation to the maximum.

The nub of the problem is that TACs are the only means by which relative stability, the fundamental principle of the CFP, can be implemented. All Member States, except Spain and Portugal, want to retain this principle. Consequently, rather than question whether TACs can achieve the objective of effectively managing multispecies, mixed demersal fisheries, there is a frantic search for solutions which retain TACs rather than to admit the conclusion that TACs are a fatally flawed system for managing such fisheries and that another method is needed. In Chapter 12 it is argued that this method should be a limitation of fishing effort.

Chapter 10

Improving the Decision-taking Process

or Can Horse Trading be Eliminated?

The background

Resolving the differences between 12 Member States and the often conflicting interests within them unavoidably requires compromise and horse trading. Compromises are reached to resolve immediate political problems. Inevitably the solution agreed is less than optimal. This chapter examines whether much of the horse trading which occurs in fisheries could be eliminated, particularly in implementing the conservation policy.

Problems peculiar to the conservation policy

The problem is particularly acute for the conservation policy because conservation measures have always been based upon scientific advice. This reflects the history of fisheries conservation. Fisheries research was started at the end of the nineteenth century in order to find the answers to complaints from the industry about falling catch rates and catches consisting predominantly of small fish. Over the years, the present system of international co-operation leading to the production of agreed scientific reports containing specific recommendations gradually evolved. These reports provided politicians and administrators with a firm basis for taking their decisions.

In contrast, the fishing industry had no similar organization. All that its members could do was to present their own ideas, based upon their own observations. Because the catching sector of the industry has so many different and opposing interests, the politicians and administrators were confronted with a series of conflicting demands. In particular, the industry did not and still does not produce a cogently argued report similar to that of ACFM. Faced with the choice between scientific certainty and practical dissension, the decision-takers inevitably chose the former. The industry became increasingly alienated as it saw that the decisions which were being taken were based not upon practical experience but scientific theory, especially as there was no forum in which these differences could be debated and reconciled. In consequence, the catching sector of the industry considers that its only option is to lobby at the highest possible level if it is to influence decisions.

Other policies

These problems are far less acute for the other policies, mainly because they are not based primarily upon scientific advice. Proposals in the area of the markets and structural policies are subject to much consultation with the fishing industry because this is the major source of information and advice available to the Commission. Additionally, the policies provide subsidies to the industry so there is a financial interest in getting proposals adopted. It is notable that the recent proposals for fleet reduction under the structural policy, which are based on the scientific advice, are meeting the same hostile response as proposals on conservation policy, even though finance for decommissioning is available. The policy for international fisheries relations, as has already been noted, is driven by the Member States, provides financial support to their fishermen and also does not suffer from the problems described.

The problems with the present system

The evolution of the system

The present system is like much in the field of fisheries, something which has evolved in a haphazard manner and which has never been critically reviewed to determine whether it responds to the needs of managing a multi-million pound industry. The system developed in the early days of the international fisheries commissions. Fisheries management was, and still is for many, implementing the scientific advice, even given formal recognition in the name of the Advisory Committee on Fishery Management of ICES. Scientific advice was formulated on the basis of biological models, largely in isolation from political influence, and presented to the fishery commissions which decided whether and to what extent it should be implemented. The decisions taken were influenced by pressures from the fishing industry, which attended the meetings. Except that the fishing industry is not allowed to attend the meetings of the Council although it is present in the corridors, and that decisions taken by the Council are binding in all Member States and not subject to an objection procedure, very little has changed.

The present system

The Community continued a system which already existed. Regulation No. 170/83 provided for the establishment of the STCF whose remit was to produce annual reports on the state of the fish stocks in Community waters. Not only had the STCF to provide advice but Article 2 of Regulation No. 170/83 required the Commission to base its proposals for conservation regulations upon that advice, in particular the report of the STCF. As described in Chapter 6, the Council has refused to adopt proposals which lacked this scientific basis.

This requirement has has been made less stringent in Regulation No. 3760/92 whose Article 4 provides that conservation measures:

'. . . shall be drawn up in the light of the available biological, socio-economic and technical analyses and in particular of the reports drawn up by the Committee provided for in Article 15'.

The committee provided for in Article 15 is the Scientific, Technical and Economic Committee for Fisheries (STECF) which replaces the STCF. The new title reflects the explicit account which is now to be taken of socio-economic factors in formulating conservation measures. But as the Council has effectively always taken into account all the factors mentioned in taking its decisions, it seems improbable that the change in emphasis will have major consequences to the way in which decisions are taken, if at all.

Problems in formulating the scientific advice

In formulating their advice both the STCF[1] and ACFM face a major problem which is that the Council has not fixed any objectives for the management of the fisheries. The proper role of these committees should be to advise the Commission on the biological implications of the Community's policy objectives and whether there are biological limitations on achieving them. In the absence of objectives, STCF and ACFM can base their advice upon biological criteria only.

Only fishery scientists are allowed to attend the meetings of STCF and ACFM. This is a deliberate policy in order to permit the advice to be formulated without any political pressures being applied. The scientists are aware of economic and political considerations and implicitly, sometimes explicitly take account of them. This is not a problem in itself but becomes one because there is no formal procedure for discussing the implications of the scientific advice with the industry and is a major reason for its unacceptability.

Consulting the industry

The Advisory Committee on Fisheries

In order to consult the fishing industry, the Commission has established an Advisory Committee on Fisheries which consists of representatives from all sectors of the fishing industry plus consumers. It has three sub-committees, resources, markets and structures. The full committee meets infrequently. The sub-committees usually meet once a year, the meetings normally lasting only one day. The frequency and duration of the meetings are determined by the small size of the committee's budget. The sub-committees have become little more than a forum for expressing national viewpoints, not even sectoral viewpoints as intended.

Relations between the Commission and the sub-committee on resources are particularly strained for the reason that the sub-committee is frequently not consulted on proposals until after they have been adopted by the Commission. This situation invariably arises over the proposal for the annual regulation fixing TACs and quotas. However, the timing of this regulation is dictated by the heavy over-exploitation of the majority of fish stocks, which necessitates leaving the assessments to the last possible moment in order to have the most up-to-date information on recruitment. Consequently, the scientific advice is not available until mid-November for the TACs which will be adopted by the middle of the following December. During this brief period several rounds of negotiations with the Faroe Islands, Norway and Sweden have to take place before the proposals can be finally formulated, leaving no time in which to consult the industry. An obvious solution would be to start the TAC a year later. However, the industry always insists on having the advice based upon the latest data. As data are collected continuously, this solution would not remedy the situation. The only solution is to stop exploiting the stocks so heavily.

The fragmentation of the industry

A major factor which makes consultation with the fishing industry of the Community very difficult is that the industry is very fragmented. This makes it very difficult to convene a representative group. Not only are there differences of view between Member States but within each Member State. This contrasts with the scientists who have few problems in co-operating and usually, but not always, find no difficulty in agreeing upon their recommendations.

This fragmentation has consequences for the negotiating procedure. More often than not, the 'official view' of a Member State represents a compromise between the differing interests of each of the sectors in its own industry. The difference of views between the National Federation of Fishermen's Organisations (NFFO) and the Scottish Fishermen's Federation (SFF) in the UK is well known and even each of these organization's view represents a compromise between those of their constituent organizations. This is hardly surprising; the majority of proposals for regulations, if adopted, would inevitably result in changes in the established practices of fishing and, even more importantly, would result in financial loss, or at least, predicted financial loss. For both of these reasons it is inevitable that one or other sector of the fishing industry opposes one or more parts of a proposal; often one group of Member States will be in favour of part of a proposal to which another group is vehemently opposed.

Taking decisions

'Reserving one's position'

In such a situation, the Member States do not try to reach agreement at the level of the working group but reserve their positions until the meeting of ministers for two reasons. First, making concessions at the level of the working group, or even COREPER, is interpreted by their fishing industries as political weakness, an accusation which no politician can afford. Second, by reserving positions until the meeting of ministers each Member State hopes that it will be able to make trade-offs with other Member States so that its position can be maintained. If its votes are perceived necessary to obtain a qualified majority a Member State can often obtain significant concessions to its position.

The proposals of the Commission are based upon the scientific advice. The proposals are then modified in the Council according to the political exigencies at the time. In this context 'political' is synonymous with 'pressures from the fishing industry'. At the stage when decisions are finally taken, during the meetings of the ministers, there is no independent scientific input. Those scientists present at these meetings–and few Member States have scientists in their delegations–are present in their role as national scientists advising their ministers, even though they may also be members of ACFM and have formulated the advice which is now being modified. The only advice which can be considered to be neutral is that from the scientifically qualified staff of the Commission, the Commission not having a national position to support. The negotiations are little more than a system of horse trading, as described in Chapters 2, 4, 5 and 6. Why does such political horse trading occur and is it justified? This chapter examines these questions, analysing the defects inherent in the present system and then proposing how it should be improved. The other policies suffer to some extent from the same defects but the dependency of the conservation policy on scientific advice creates most of the problems described.

Is horse trading justified?

Given the conflicting interests of Member States, compromises are inevitable if agreement is to be reached. But it would be far better if decisions were taken as a result of a process of calm evaluation of what is best for the industry in the long term, not of what is expedient in the short term to resolve immediate political problems. Under the present system the level at which this could best occur is that of the Council working groups where there is most expertise on these matters. Also, the working group is usually not operating under time constraints and there is ample opportunity for the national representatives to consult their fishing industries between meetings. However, for the reasons already described, the working groups are not allowed to make modifications; ministers, under pressure from their fishing industries to make no concessions, reserve their positions until the meeting of the Council. There they take decisions on the advice of their senior administrators who have no practical knowledge of fishing whatsoever and, under the conditions prevailing, have little or no opportunity to consult with representatives of the industry.

This is particularly true if the meeting goes into restricted session. It is not unknown for these sessions to end with the ministers emerging to state that agreement has been reached but unable to provide their officials with the details of what has been agreed, which is vital to drafting the final legislation.

All then depends upon the secretariat of the Council which, very efficient as it is, is not technically qualified. Additionally, given the pressure under which proposals must be drafted, often very late at night or, more usually in the early hours of the morning, sometimes after 24 hours or more of non-stop negotiation, it is inevitable that conflicts which arise between different parts of the legislation are not noted at the time. Even if they are detected they may be ignored because of the necessity to present the draft compromise to the ministers without delay. It is often only in the cold light of morning that the mistakes are discovered, by which time it is too late because the proposal has been adopted and is Community law.

Much of the fishing industry may criticize the legislation which is adopted as being complex, often contradictory, difficult to implement and impossible to enforce. These criticisms are often valid but the legislation reflects the political pressures which the fishing industry of each Member State places upon its ministers not to concede at all, or if concessions must be made, to concede only at the last possible moment at the highest possible price.

Horse trading is an inevitable result of the organization of the Community and the manner in which the negotiations between the Council and the Commission take place. It can be claimed that horse trading is justified because it is the only means by which to make the present system work but it is highly unsatisfactory and leads to decisions which are far from the best interests of the fishing industry. In such a situation, it would be far better to improve the system.

Towards a better system

The present situation may have been satisfactory in the era of the international fisheries commissions when there was an objective, that of F_{max}, which was implicitly if not explicitly accepted, and there was little that could be done except to adopt technical conservation measures, mainly minimum mesh sizes. It is inappropriate for today's conditions and will become increasingly inappropriate if the Council is to establish management objectives and management strategies which are to take into account 'appropriate social and economic conditions', as provided by Regulation No. 3760/92. It is also totally illogical that input from the fishing industry should be left until the final point of decision taking when that input has so much influence on the final outcome and often results in perverse decisions as a result of its timing and the way in which it has to be taken into account.

Two fundamental questions need to be examined. Is the means by which scientific advice is now provided satisfactory and how can the manner in which the fishing industry provides an input into management be improved?

The provision of scientific advice

The present role of the Scientific and Technical Committee for Fisheries

A knowledge of the fish stocks and their responses to fishing is essential for fisheries management and scientific advice is therefore vitally important for fisheries management. The 'official source' of scientific advice to the Commission, as provided for in Article 2 of Regulation No. 170/83, was the STCF and will continue to be for the STECF under the provisions of Article 4, paragraph 1, of Regulation No. 3760/92. However, in practice, the main source of advice is the Advisory Committee on Fishery Management of ICES. One reason for this is that ICES has developed over many years both the organization and computer resources for making assessments. Also, many of the stocks concerned are joint stocks, shared with Norway, which cannot be a member of the STECF and for which the advice of ICES is the definitive advice. Also, the Framework Fisheries Agreements with the Faroe Islands, Norway and Sweden provide that negotiations shall be based on the advice of ICES. Management of the stocks in the Baltic Sea is undertaken by the International Baltic Sea Fisheries Commission to which ICES is also the official provider of scientific advice. ICES had established a dominant role decades before the STCF was created so there was no requirement for the latter to make its own assessments nor for the Commission to provide the STCF with the necessary resources. Even if the Commission had attempted to set up a parallel organization, there are reasons for believing that it would have failed. It would have required a substantial budget which Member States would probably not have been prepared to provide because most of them already fund ICES. Perhaps more importantly, it would have meant having to provide the Commission with all the basic scientific data to which it would then have had access; national data are held at ICES but on confidential files to which access is strictly limited.

In consequence, the STCF plays a very limited role in providing advice. Its annual meeting is short, usually lasting only five days during which it makes few, if any, assessments, not having the necessary facilities. Attempts to extend the meetings would be resisted by the Member States because most members are government scientists whose prime responsibility is to their ministries. The scientists themselves do not want to spend time away from their research programmes on work which they consider of little value. This is aggravated by the fact that membership is unpaid and allowances do not cover expenses. In addition to the annual meetings, the Commission convenes the STCF to meet from time to time in order to advise it on particular issues. As these meetings are often held at short notice and without adequate or any background information being provided in advance, the ability of the STCF to respond is very limited. The Commission has attempted to enlarge the role of the STCF by establishing working groups which examine long-term issues, such as managing the multispecies, multifleet fisheries of the North Sea, taking into account economic as well as biological data. This work is carried out in national laboratories, effectively becoming part of national research pro-grammes, with co-ordination at occasional meetings of the working group. The present situation is very unsatisfactory and totally inappropriate for providing the advice on decisions which determine the operations of the fishing indus-try. How might it be improved?

ACFM or STECF?

The present duplication of work is wasteful of resources but is ACFM the appropriate committee to provide advice to the Commission? If it is, should the STECF be abolished? The main argument for ACFM continuing to be the advisory body is that it has an established system, the databases and the com-puting resources. But there are several serious drawbacks to its continued use. A major disadvantage is that, although the Commission has an Agreement of Cooperation with ICES, under which ICES accepts to respond to questions on fisheries management, the Commission is entirely in the hands of ICES regard-ing the timescale and the way in which it responds. There is no contractor-customer relationship which permits any debate between the Commission and ICES. To obtain effective management advice the Commission needs an advi-sory committee which has to respond to its requirements within an agreed time limit and which permits full and open discussion of the issues.

An equally important question is whether ICES can provide advice which has an economic as well as a biological basis. There has been much discussion of this issue within ICES which has concluded that it cannot draw on exper-tise in this field. More importantly, the economic criteria for decision taking differ between its contracting parties. An economic input is essential to fish-eries management.

A third question which must be asked is whether the method of nominating the members of ACFM is appropriate to its responsibilities, even under pre-sent circumstances? The meetings of ACFM are restricted to its members, the ICES statistician and, in recent years, two scientifically qualified observers, one from the Commission and the other from the Faroe Islands. Except for the chairmen of the ICES Consultative, Demersal Fish, Pelagic Fish and Baltic Fish Committees, membership is on a national basis, one from each contracting party to ICES, candidates being selected by the national authorities. The nom-inated members are automatically elected by ICES but this is a formality. The chairman of ACFM is elected by its members but has to be confirmed by the Council of ICES. This system means that the membership does not consist of the most able stock assessment scientists. Furthermore, the manner in which members are nominated is that they are all scientists from national fisheries laboratories, which inevitably means that their common outlook will have been formed by operating in an environment of providing advice to their national governments and will have been further formed by operating within the framework of ICES. Fisheries experts from non-government organizations who might have different attitudes are excluded, not only from ACFM but from the meetings of ICES, except by invitation, making ICES a closed society.

Like the members of STCF, those of ACFM all have responsibilities to their national laboratories, the work of ACFM being additional to their normal duties. The meetings of ACFM are of longer duration than those of STCF, there being two meetings a year of approximately ten days each, but these are inadequate for the work involved, even though the major part of the assessment work is done by the various working groups prior to its meetings.

It must be concluded that ACFM is not an appropriate committee for providing advice to the Commission in the framework of a proposed fisheries management system.

A revised role for the STECF

How might these problems be resolved? Fisheries management has to be based upon the biological state of the stocks and advice on the stock situation is an essential element of management. However, it is only one element in taking management decisions which must depend ultimately upon the economic and social objectives which the Community wants to achieve. The STECF must have expertise in fish stock assessment, gear technology, fleet management and economics in order to do its job effectively and to fulfil the remit laid down in Article 16 of Regulation No. 3760/92. To function effectively and efficiently the Commission must be able to convene the STECF when required, present it with questions and set the timetable for a response. The present organization of the STECF does not permit this. These requirements call for a permanent committee responsible to the Commission. Given the fundamental role of this committee and its enlarged responsibilities under Regulation No. 3760/92, members of the STECF should be full-time, paid by the Commission, with membership being of limited duration, say three to five years; the task is too important for it to remain a part-time, unpaid occupation as it is now. However, it would be counter-productive to make members full-time civil servants of the Commission, based in Brussels. Within a short period they would have lost contact with on-going research but, perhaps more importantly, this would almost certainly deter the best qualified persons from becoming members. There is no reason why the members should not remain located in their original institutions to which right of return would have to be guaranteed. Computer links, facsimile machines and other electronic equipment would permit the members to work at separate locations but exchange data and information, with meetings being held when and where necessary. This would also avoid domestic problems of transferring staff to Brussels.

In many respects the proposed staffing arrangement would be no different from that which exists already because many members of ACFM, in particular its chairman and the chairman of the Consultative Committee, do work almost full time for the committee, except that they are unpaid and nominally still work for their national laboratories.

The role of the fishing industry

This is the most intractable problem to resolve. One of the problems is that the catching sector of the fishing industry is divided into many different sectors, each with its own interests which often conflict with those of other sectors. Not only is nationalism rife but within countries different, conflicting views exist. The activities of those fishing with fixed gear conflict with those who use mobile gear. The interests of those who fish for species which require nets having a small mesh size conflict with those who fish for species for which it is better to use a large mesh size. National fisheries organizations representing the industry attempt to co-ordinate these different views but they are constrained by the fact that their primary responsibility is to represent the interests of their members. Consequently, their views can represent only the lowest common denominator which least affects the majority. Nowhere is this diversity of interests better illustrated than in the UK where there are two major national, umbrella organizations, the National Federation of Fishermen's Organisations, covering England and Wales, and the Scottish Fishermen's Federation covering Scotland, 15 producers' organizations and at least 66 local associations covering specific interests, often at port level. Consultation is further complicated by the fact that there are three government departments responsible for fisheries, the Ministry of Agriculture Fisheries and Food (for England and Wales), the Scottish Office Department for Agriculture and Fisheries, and the Department of Agriculture for Northern Ireland in addition to 12 Sea Fisheries Committees regulating fisheries within the three-mile coastal zone in England and Wales. Given these conflicting interests and their exclusion from the decision-making process at an early stage, each sector of the fishing industry has become habituated to pressing its own case at whatever political level it can obtain a hearing. Politicians are forced, as a result, to take decisions concerning conflicting interests which please no-one.

A Fisheries Management Committee

One possible means of resolving this problem would be to make representatives of the fishing industry members of the STECF, reformulated as suggested. But the essential role of the industry is to provide an industry input into the decision-taking process whereas that of the STECF should be to examine how the policy adopted by the Council could be implemented and the limitations to implementing it. It is more appropriate to consider how the role of the Advisory Committee on Fisheries could be developed. The committee could then be reformed as a Fisheries Management Committee whose role would be to provide a 'political' input into the decision-taking process, replacing pressure from the fishing industry on ministers as occurs under the present system.

The committee would consist of active members of the fishing industry. This implies that they should not become full-time committee members, as proposed for the STECF, but should continue with their existing work, being fully financially compensated for all time spent on the work of the committee. Representation should be on a sectoral basis as at present, with the obligation to represent the interests of the sector, not national interests as now occurs. To enable it to function efficiently the committee should be small and be provided with its own secretariat. It should also have its own budget in order that financial restrictions should not limit its activities, as they do of the Advisory Committee on Fisheries. (How sufficient finance could be provided is suggested in Chapter 12.) For example, it should be able to employ its own experts to evaluate particular strategies. The Fisheries Management Committee would be required to operate closely with the STECF, which would provide it with the results of its activities. These would then be discussed with the Fisheries Management Committee, which would then submit to the Commission proposals for draft legislation.

The formal decision-taking process

How would such a system operate within the decision-taking processes of the Community? The suggested structure is shown in Fig. 10.1. To make this structure work efficiently, it would be essential to adopt a policy. Neither the STECF nor the proposed Fisheries Management Committee can operate effectively if the objectives are unknown. Also, the Council would have to revise its political appreciation of the importance of fisheries and act accordingly. Fisheries now have a high political profile out of all proportion to their relative importance in terms of gross national product (GNP). In most Member States fisheries constitute less than 1% of GNP. The objective of establishing the new structure would be to resolve all the 'political' problems arising from the different viewpoints of the industry within the Fisheries Management Committee whose reports would form the basis of proposals from the Commission. These proposals would become little more than a formality to be adopted as 'A' points.

A major problem would be initiating the system because its function depends upon having an agreed policy which the structure would foresee being determined in conjunction with the STECF and the Fisheries Management Committee. But there is already almost an agreed policy which is that the size of the fishing industry should be reduced, multi-annual guidance programmes being the means by which this is to be achieved, although the objective is to conserve the stocks, not economic fisheries management. The starting point could be for the Council formally to adopt as an agreed policy a specific reduction in fleet capacity as a target and to decide in the interim the basis on which it will fix TACs. The Commission could then ask the two committees to advise it on the implications of all the issues which this would raise, although the extent of the reduction has not been determined.

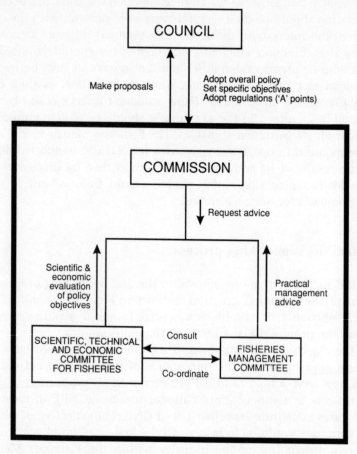

Fig. 10.1 Proposed organization for the provision of advice on fisheries management (see text for further explanation).

Summary

The present system of decision taking in the Community is totally inappropriate for implementing an effective conservation policy. There are historical reasons why the system has developed as it has but the situation has changed and the system needs to be changed accordingly.

There are two major deficiencies of the present system. First, advice on management has to be formulated in the total absence of an agreed policy. Second, the fishing industry is totally excluded from the formulation of management advice. The result of this latter deficiency is that the industry lobbies at the only level at which it can influence decisions, that of ministers. Meetings of the Council are not the appropriate place to take decisions concerning technical matters. Management decisions should be based upon considered evaluation of the facts and the possible options, not on compromise and horse trading.

A better system of providing advice is needed, one that is based upon the STECF having a similar role to that which it has now but with full-time members, and upon an effective Fisheries Management Committee, drawn from the industry, which has an effective role to play.

Note

(1) The acronym 'STCF' is used when reference is primarily to the past role of the STCF. The acronym 'STECF' is used when reference is specifically to the role of this new committee.

Chapter 11

Towards a New Policy: Setting the Objectives

or Taking Difficult Decisions

The need for objectives

The lack of specific objectives

There is an implicit assumption in the way in which fisheries are now managed that there is no need to set specific objectives. It is assumed that the objectives are automatically dictated by the biological factors which determine the relationship between catches and rates of fishing and that fisheries would be properly managed if the recommendations made by fishery scientists were implemented. It is not surprising that this view exists. Over-fishing was recognized at the end of the 19th century but its causes could not be identified until the necessary research had been carried out. The recommendations for management were based upon the results of this research. Fishery management became synonymous with applying the scientific recommendations.

This viewpoint was reinforced in the period between 1946 and 1977, during which most fishing took place in international waters. Management of the fish stocks in these waters became the responsibility of international fishery commissions. The contracting parties to these commissions had no common ground on which to base the management of the resources other than the scientific advice. Capitalist countries were fundamentally interested in maximizing profits. Communist countries were interested in maximizing catches. All countries were competing for the same resource and the only way in which each considered that it could achieve its national objectives was to catch as much fish as possible. However, they had all ratified the conventions under which they were contracting parties to the commissions and they were therefore under an obligation to implement the objectives of the commissions, 'to promote the conservation and optimum utilization of the fishery resources', to use the phraseology of the Convention on Future Multilateral Co-operation in North-east Atlantic Fisheries. The only common denominator upon which the contracting parties could agree was to implement the scientific advice which, at this period, was to achieve the maximum yield per recruit (see Chapter 8). Another factor played an important part in reinforcing this attitude. While politicians and administrators could not agree on management objectives fishery scientists from the contracting parties could co-operate and agree on recommendations, producing reports which provided a firm basis on which the administrators could base their decisions.

The objectives of the CFP, 1970–92

When the first steps towards establishing the CFP were taken in 1970 with the adoption of the markets and structural policies, fishery limits were still a maximum of only 12 miles in any of the Member States. Even if the Council had considered it necessary to adopt specific objectives, the limited size of national fishery limits meant that they could not have been achieved. The Council did adopt general objectives which were 'to promote harmonious and balanced development of this industry within the general economy and to encourage rational use of the biological resources of the sea and of inland waters'.

By the time that the conservation policy was adopted in 1983, 200-mile fishery limits had been in place for six years. The establishment of these limits provided the opportunity to manage the fisheries to achieve specific objectives, but the opportunity was not grasped. The objectives, as provided by Article 1 of Regulation No. 170/83 were:

> 'to ensure the protection of fishing grounds, the conservation of the biological resources of the sea and their balanced exploitation on a lasting basis and in appropriate economic and social conditions'.

This text embraced biological, economic and social objectives but provided no criteria by which the balance between them was to be decided. There were two reasons for this. First, each of the Member States had its own aspirations and they were no more able to agree on specific objectives than the contracting parties to the fishery commissions. Second, the administrators who negotiated Regulation No. 170/83 were those who had represented their countries in the fishery commissions. Their mentality was conditioned by what had happened in those commissions and they saw no reason to change their outlook. The voices of those few who suggested that a different approach could now be tried because 200-mile fishery limits had been adopted and that there was a 'Community sea' were ignored.

The objectives of the CFP, 1993–2002

The review of the conservation policy in 1992 provided the chance to determine specific objectives but those in Article 2 of Regulation No. 3760/92 are no more specific than those in Regulation No. 170/83. They are:

> 'As concerns the exploitation activities the general objectives of the common fisheries policy shall be to protect and conserve available and accessible living marine aquatic resources, and to provide for rational and responsible exploitation on a sustainable basis, in appropriate economic and social conditions for the sector, taking account of its implications for the marine ecosystem, and in particular taking account of the needs of both producers and consumers.'

These objectives expand on those stated in Regulation No. 170/83 but they provide no better basis for effective management. Words such as 'rational' and 'responsible' beg too many questions, especially when linked with the need to take into account appropriate economic and social conditions, implications for the marine ecosystem and the needs of both producers and consumers.

Article 8, paragraph 3 of the same regulation also refers to objectives, stating that:

'The Council . . . may establish management objectives, on a multiannual basis, for each fishery or group of fisheries in relation to the specific nature of the resources concerned. Where appropriate these shall be established on a multispecies basis. Priority objectives shall be specified including, as appropriate, the level of resources, forms of production, activities and yields.'

The same paragraph provides that for fisheries for which management objectives have been set, management strategies shall be established.

The implications of this text are interesting, to say the least. That 'the Council *may* establish management objectives' suggests that the Community does not consider setting management objectives for fisheries is an essential requirement but something to be undertaken only when circumstances, unspecified in the regulation, so require. That the regulation considers it necessary to provide that they shall be on 'a multiannual basis' is also curious; how can fisheries be managed otherwise? (This is almost certainly a coded reference to multiannual TACs.) No guidance is given as to how the Council is to decide between the various priority objectives, objectives which are inter-related and in one case, the level of the resources, outside the control of the Council. It will be very interesting to see how the Council implements these provisions, if at all.

Some clues as to the objectives which might be adopted are provided by the 'Report 1991 from the Commission to the Council and the European Parliament on the Common Fisheries Policy' (referred to subsequently as the 'Report 1991') which the Commission had to produce under the provisions of Regulation No. 170/83. It formed the basis for the review of the conservation policy which took place in 1992. In this report, the Commission suggested (this word is used to distinguish ideas in this report from formal proposals of the Commission) specific objectives, which were as follows:

- to ensure the sustainability of the fishing industry by making the necessary cuts in fishing effort, including fishing capacity;
- to resolve the social problems created by the reduction in capacity;
- to guarantee stable supplies at reasonable prices to the consumer;
- to contribute to economic and social cohesion.

It is more than possible that if and when the Council does establish management objectives, it will base its decisions upon these suggestions, especially as the Council has to act on a proposal from the Commission, which will automatically build upon its original ideas. While these objectives have the advantage of being much more specific than those contained in Regulation No. 3760/92, their precise interpretation is uncertain. Also, they have been to some extent surpassed by events. When the report was written the Commission was attempting to obtain additional funds for achieving the second objective, that of resolving social problems. As described in Chapter 2, this would have been Objective VI of the European Structural Funds. Even though these objectives had not been adopted, the Commission considered that restructuring was unlikely to be successful unless finance for resolving the social problems was made available. In this, the Commission is almost certainly correct. The fourth objective implied that most of any funds obtained would have gone to the poorer Member States. However, this report was written prior to the European Summit of December 1992 where events resulted in the attempt to establish an Objective VI in the Structural Funds having to be abandoned (see Chapter 2).

The main problem lies with the first objective suggested by the Commission, that of ensuring the sustainability of the industry. The general tenor of the report is that fisheries should continue to be managed essentially on a biological basis in order to maintain a balance between fishing capacity and resources. The implication is that this balance should be that of F_{max} and that multi-annual guidance programmes are to be the means by which capacity is limited. While the report calls for more research on economic issues, it does not examine how the fisheries are to be made more profitable, it apparently being assumed that the envisaged reduction in the size of the fleets will automatically result in this happening. But as the fishing industry could be sustained at a very low level of profitability, or none at all if it continues to be subsidized through the structural and markets policies, it can be imagined that the Council will decide that the 'necessary cuts' should be kept very small in order to minimize both the difficulties of social problems and the budgetary requirements arising from a major restructuring of the industry.

The provisions of Regulation No. 3760/92 show that there has been no change in outlook and, most critically, no specific objectives have been adopted. There is even the implication that fixing objectives is only an occasional necessity.

The objectives provided by the Treaty

It is a curious feature of the conservation policy that neither Regulation No. 170/83 nor No. 3760/92 has drawn upon, let alone incorporated the objectives laid down in Article 39 of the Treaty. These objectives were elaborated for the Common Agricultural Policy but as the CFP derives its legal basis from the provisions for the CAP, they legally apply equally to the CFP, even though they are phrased in terms which are more appropriate to agriculture than fisheries. The only occasion on which these objectives have been formally considered in terms of the CFP was in the 'Report 1991' in which the objectives suggested by the Commission were derived from those in Article 39. The objectives provided by Article 39 are:

'(a) to increase agricultural productivity by promoting technical progress and by ensuring the rational development of agricultural production and the optimum utilization of the factors of production, in particular, labour;
(b) thus to ensure a fair standard of living for the agricultural community,

in particular by increasing the individual earnings of persons engaged in agriculture;

(c) to stabilize markets;

(d) to ensure the availability of supplies;

(e) to ensure that supplies reach consumers at reasonable prices.'

Interpreting the objectives

Fishing is a different activity to agriculture. The latter is based upon the management of private property, a farm, whose inputs are directly under the control of the farmer and whose outputs are almost directly related to the inputs, although, of course, affected by weather. Unlike fisheries, the outputs are not affected by what other farmers do. In contrast, the outputs from fisheries depend very much upon what other fishermen do. Also, fishermen have no control over the initial abundance of their resource, which is almost totally out of the control of man for the reasons described in Chapter 8. Despite these fundamental differences, it is possible to interpret these objectives in terms relevant to fisheries.

Increasing productivity

It is essential to distinguish between productivity and production because the two words are very often used as if they were synonymous, which is frequently the case in the context of the CFP. Productivity is the ratio between outputs and inputs. Production is output only. In fishery terms productivity is catch per unit effort. As previously shown in Fig. 8.6, productivity in a fishery is highest when the rate of fishing is lowest and decreases continuously as fishing effort increases. Unlike farming, increasing inputs results in lower, not higher productivity. Therefore, to achieve the first objective of the CFP, as defined by the Treaty, would imply having the smallest possible fleet and fishing the stocks very lightly.

A fair standard of living

This objective begs the question of what is meant by 'fair'. Obviously, if the first objective were realized, there would be very few fishermen. All of them would be extremely rich because each would be getting a high return on his investment. What is considered 'fair' can be only a subjective decision. Making the assumption that a few fishermen earning high net incomes by excluding other potential fishermen from earning a living is not 'fair', then the second objective would be achieved by reducing productivity, although there is no objective basis for deciding at what level the standard of living becomes 'fair'. Clearly if there are too many fishermen, no-one gets a 'fair standard of living' because the sale of fish (outputs) no longer covers the cost of inputs.

Stabilization of markets

This objective could be considered to be met by ensuring that catches did not fluctuate markedly from one year to the next. Ideally, they should not fluctuate at all but this is impossible to achieve in fisheries because the abundance of the raw material, fish, fluctuates naturally from one year to the next as explained in Chapter 8. Although man has no control over the cause of these fluctuations, they can be minimized. Stocks of fish consist of a series of year-classes, spawned in successive years, each of a different size. When a fish stock is lightly exploited, a large number of fish survive from one year to the next with the result that fish from many year-classes are present in the sea. Under these circumstances, the large year-classes offset small year-classes and the greatest stability is ensured. As the rate of fishing decreases, the number of year-classes in the stock decreases because fewer fish survive to reach old or even middle age. A small or a big year-class now has a bigger influence on catches, which start to fluctuate markedly in line with the size of the most recent year-classes. When rates of fishing become very high, catches depend on only one or two year-classes and, if the variation in year-class size is very large, swing violently from highs one year to lows the next. This is the situation today for North Sea and west of Scotland cod and haddock (Fig. 7.1) as well as many other stocks. This leads to the conclusion that this objective can be met only if the stocks are not fished too heavily. However, stability cannot be guaranteed; if there is a long series of consecutive small year-classes, catches will fall; that is inevitable.

Ensuring availability of supplies
In the context of the conservation policy, this objective may be equated with ensuring that fish stocks are not so heavily exploited that they collapse, with the consequent closure of fisheries. In the broader context of the CFP, ensuring availability of supplies is also achieved through agreements with third countries and trading agreements concerning fish.

Reasonable prices for consumers
This also begs a question, that of what is meant by 'reasonable', but assuming that the average price of fish will be lowest when total landings are highest, the consumer will obtain the cheapest supplies when landings are highest. This begs many questions about imports and whether the term 'reasonable' also has implications for the catching and processing sectors.

Setting objectives

The conflict between different objectives

As shown by the discussion on the objectives laid down by the Treaty, it is impossible to manage fisheries in a way which will achieve all the objectives simultaneously. Unfortunately, this crucial issue has always been avoided and decisions taken as if optimizing all the objectives were possible. It is this thinking which explains the vagueness of the objectives given in Regulations No. 170/83 and No. 3760/92. It is also this thinking that leads the fishing industry to believe that fishing can give employment to large numbers of fishermen who, at the same time, can be ensured of a reasonable income. For example, it is commonplace for fishermen to argue that their sons must be able to follow father into the industry, ignoring the fact that this must mean that the total catches must be shared out amongst more fishermen. With more fishermen, the total amount of fish available to share between them will be less, not more, for the reasons described in Chapter 8.

Even if it proved possible to agree on an overall objective, there would still be conflicts to resolve about sub-objectives. As has been stated many times, it is impossible to optimize the harvesting of all the species in the sea. Even on an issue where there is general agreement, that priority should be given to catching fish for human consumption, which fisheries should be optimized where there is a conflict of interests; to take a few examples, should it be the fisheries for haddock and whiting or those for cod and saithe, or those for *Nephrops* or hake, or those for sole or plaice? In each example, the optimal exploitation of the last named species would require a very large minimum mesh size and result in insignificant catches of the first named. This is basically a simple issue but difficult to resolve because two different groups of fishermen are involved with the gains accruing to one group and the losses being sustained by the other. It is the failure to address these problems and to fix specific objectives that has resulted in the regulations concerning minimum mesh sizes being so complex.

Some possible objectives

What should be the overall objective? This section examines some of the possibilities and their advantages and disadvantages.

The objectives are based upon the biological model which describes how a fish stock reacts to fishing, as described in Chapter 8. The model provides a basis for determining a management objective but offers no basis for how that objective is to be achieved in practice. The situation is further complicated by the fact that, as also described in Chapter 8, the shape of the yield per recruit curve and any selected point upon it, such as F_{max}, is determined by the rates of fishing mortality on each age group. No yield per recruit curve is unique and no reference point is unique. All can be infinitely varied. Therefore an objective cannot be fixed in isolation. All the factors affecting the fixing and achieving of an objective must be considered simultaneously.

Maximizing employment

The political imperative to maintain employment is reflected in the many references in the legislation to operating the conservation policy in accordance with 'appropriate economic and social conditions'. In practice this has resulted in the failure, in general, of the Council to adopt technical conservation measures which could be effectively enforced and TACs which corresponded with reductions in rates of fishing. One of the reasons for the failure of most Member States to meet the targets fixed by the multi-annual guidance programmes has been the impact of these programmes on employment.

There are two ways in which high employment can be maintained. Either a large number of fishermen can be permitted to fish without restriction or the fishing possibilities shared out between them by some means, for example, by TACs or limiting the number of days at sea. The first scenario implies unrestricted fishing and its outcome has already been examined; it is low and highly variable total catches, very low catch rates and very low incomes. This scenario is incompatible with having a CFP.

The second scenario implies limiting the rate of fishing and thus reducing total fishing effort in order to achieve a specific objective, such as the maximum sustainable yield (MSY). Its implications are more complex. Reducing the amount of time which each fisherman is allowed to fish should result in higher catch rates. However, individual fishermen will be financially no better off unless the increase in catch rates compensates for the time that they are no longer able to fish. For example, to compensate for a reduction in the amount of fishing by 50%, catch rates would nearly have to double. (They would not have to double because there would be gains from the reduction in variable costs, such as fuel.)

There is a major problem in implementing this scenario, well illustrated by what has happened in the case of TACs. It would apply equally to attempts to reduce the amount of fishing by limitations on days at sea. Assuming average recruitment, there is a time-lag of two or more years between reducing the amount of fishing and catch rates increasing. During this period the reductions in fishing time would result in a fall in incomes. In order to compensate, fishermen would do their utmost to increase efficiency, by using greater quantities of gear if possible, such as more traps or greater lengths of gillnets, or by fishing more hours per day at sea, if possible. To achieve the objective requires that severe constraints would have to be placed upon the activities of the catching sector and that these restraints were effectively enforced.

There is also a fundamental economic objection to reducing fishing effort in this way. The amount of capital employed is not reduced and therefore the costs of servicing it remain unchanged. The profitability of the fishery is increased only by the amount of additional output and the savings on variable costs. Fixed costs remain unchanged. As a social objective, maintaining employment may have much to recommend it. In economic terms it is not a sensible objective.

Maximizing catches
Maximizing catches essentially implies implementing the concept of the maximum sustainable yield (MSY). But fixing MSY as an objective does not determine how the fishing mortality rate at which it is taken is to be achieved. It could be achieved by having a fleet which was limited to that which, permitted to fish without restriction, generated only the required rate of fishing. Alternatively, it could be achieved by having a large fleet and a large number of fishermen but restricting their activities as already described.

Maximizing catches does not necessarily result in maximizing gross revenue, although the catching sector of the industry acts as if it does, always arguing for the highest possible TACs. If price is very sensitive to supply, the highest gross revenue may be provided when catches are much less than the maximum.

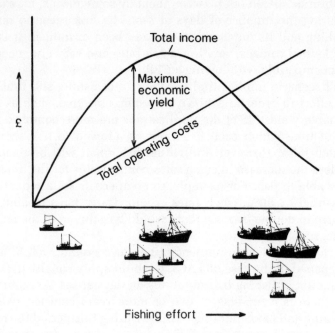

Fig. 11.1 Yield per recruit curve showing point of maximum economic yield.

Setting MSY as the objective can be considered in isolation only if reasonable prices for consumers is taken as the overriding objective, irrespective of how this is achieved.

Maximum profitability
The maximum profit from a fishery is obtained by exploiting the stocks at a point at which the difference between the costs of fishing and the income from the sale of fish caught is at a maximum. This determines the maximum economic yield (MEY) and is demonstrated in simple terms in Fig. 11.1. Implicit in the concept of MEY is that fishing capacity is limited to that at which, if vessels were permitted to fish freely, they could not generate more than the rate of fishing mortality corresponding with the MEY. Otherwise the economic benefits are dissipated.

The objective of MEY is that which provides the best compromise in meeting the objectives laid down by Article 39 of the Treaty. Productivity would be high and those fishermen who were permitted to fish would have a fair standard of living. Markets would be much more stable than they are now, although less so than at lower rates of fishing. Supplies would be ensured because it is improbable that stocks would collapse at rates of fishing corresponding to MEY, certainly not as a result of fishing. Collapses consequent on environmental changes can never be excluded. Finally, total catches although less than the theoretical maximum, would be only slightly less, thus ensuring reasonable supplies for consumers.

Biological objectives

There exist several parameters which are described as 'biological reference points'. These include MSY and $F_{0.1}$. The concept of $F_{0.1}$ was originally derived as an economic objective, similar to MEY but implying a reduction in the rate of fishing less that required by MEY. The concept has been taken over as a biological reference point but it is 'biological' only in the sense that fishing at this level poses less risk of the stocks collapsing and also less risk of over-fishing, in the strict sense of fishing at a rate higher than that at which the MSY will be obtained. There is as much justification to call MEY as $F_{0.1}$ a biological reference point. Both are objectives but both require specific measures to be taken in order to determine how they are to be achieved.

The only specific biological objective is that of maintaining a minimum spawning stock biomass. That there is no obvious relationship between spawning stock biomass and recruitment makes it difficult, if not impossible, to determine with any precision minimum spawning stock biomass. Any figure must be little more than arbitrary, selected upon the basis of the precautionary principle. However, there is a fundamental objection to using the maintenance of a minimum spawning stock biomass as the management objective which is that the spawning stock biomasses fluctuate from year to year depending upon the size of year-classes. As a primary objective it would mean that very high rates of exploitation could be permitted when recruitment to the spawning stock biomass was high but would have to be drastically reduced or fishing stopped, when recruitment to the spawning stock biomass was low. As described in Chapter 6 this situation has already arisen in the fixing of TACs for North Sea plaice. In practical terms this is a totally unsatisfactory basis for management as it would mean changing the activities of the fleets from year to year and takes no account of the economics of the industry.

This is not to say that maintaining the size of the spawning stock biomass is unimportant. But it should be achieved through means which take into account the economic activities of the fleets.

Who should determine the objectives?

The role of politicians

Deciding management objectives should be the responsibility of the politicians. This is specifically recognized in Article 8 of Regulation No. 3760/92 which provides for the Council to establish objectives, as already described. To date they have failed to do this. In the situation prior to 200-mile fishery limits it was not possible and since then, partly because politicians were used to operate under the conditions applying in the situation prior to 200-mile fishery limits, they continued to fail to set objectives. More importantly, politicians have been unwilling to take the difficult decisions which fixing objectives involves. They are difficult because they require making choices not only between different levels of employment and different rates of reward but also different groups of fishermen, each having specific and conflicting interests.

Difficult though the decisions may be, taking them is the responsibility of the politicians.

The role of fishery scientists

Fisheries cannot be effectively managed without understanding the biological principles which underlie fisheries management. But the biological models do not determine the management objectives. All that they do is to predict how the fish stocks will react to certain levels of fishing or patterns of exploitation. How is it then that fishery scientists appear to determine the management objectives? This situation has arisen by default. Fishery scientists are faced with a difficult situation. Their proper role should be to advise on how the objectives selected could be achieved and the biological constraints on achieving them. Instead, they are expected to operate in a vacuum. Within the Community, the Council has decided that there should be a system of technical conservation measures and TACs but failed to decide what they should achieve. In the absence of specific objectives, fishery scientists have had to develop their own guidelines as to how they should provide advice. They cannot examine every conceivable scenario but must make choices. The choices are dictated by the regulatory framework, in the case of the Community, TACs and technical conservation measures. They recognize that they are competent to provide advice that is only biologically based and the objectives which they have chosen have had inevitably to be biological objectives.

The politicians have been only too willing to accept these biological objectives, especially as the scientific advice provides them with a ready-made excuse and a scapegoat when they feel constrained to make difficult decisions. At the same time, they are prepared to ignore the scientific advice when they consider that they can appeal to unspecified socio-economic factors.

Towards a profitable future

Choosing the objective

Managing fisheries involves managing the inputs, the size of the fleets, where and how they fish because both the outputs, the size of catches, and profitability are totally dependent upon these inputs. The link between inputs and outputs in biological terms is fish stock conservation, an essential element but only one element of fisheries management. Fish stock conservation is not fisheries management, although these terms are widely believed to mean the same. Fisheries management involves setting specific objectives, establishing a long-term plan to achieve them and, having achieved them, maintaining the new situation.

The Community is based upon a capitalist philosophy. In the Community fishing is one of many economic activities whose objective is to make a profit, in this case from the catching, selling and processing of fish, the key link in

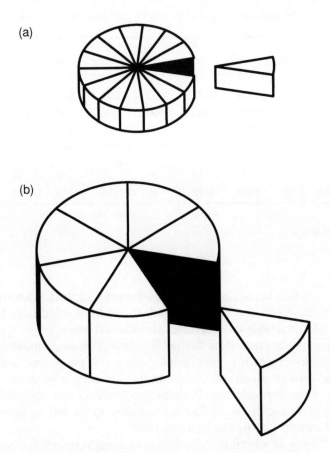

Fig. 11.2 (a) With many fishermen the 'cake' of total catch is small and the 'slice' of each fisherman is also small. (b) With fewer fishermen, the 'cake' is bigger and the 'slice' of each fisherman is much larger.

this chain being the ability to earn more from selling the fish caught than it costs to catch them. It is considered for all the reasons given in this chapter that the primary objective of fisheries management for the Community must be maximizing the economic benefits from its fish resources, essentially MEY. The CFP should be organized to achieve this objective. The essential element in achieving this is to limit the number of fishing vessels in order that total catches increase but, more importantly, those increased catches are shared between fewer fishermen in order that each may obtain a 'fair' standard of living. Both the size of the cake needs to be increased and the number of slices reduced (Fig. 11.2); compare this figure with Fig. 8.8

Quantifying the benefits

The general theory of fishing described in Chapter 8 and illustrated in Fig. 8.7 states that an unregulated fishery expands until it is no longer profitable. Although all fisheries in Community waters may not be unprofitable, the

Table 11.1 Effects on costs, earnings and wages of reducing the amount of fishing.

Reduction in the amount of fishing	0%	−10%	−20%	−30%	−40%	−50%	−60%	−70%
(1) Total income[a]	3000	3152	3273	3333	3212	3091	2848	2515
(2) Variable costs[a]	2000	1800	1600	1400	1200	1000	800	600
(3) Total wages[a]	1000	1040	1080	1100	1060	1020	940	830
(4) Profit[a]	0	312	593	833	952	1071	1108	1085
(5) Number of fishermen[b]	137	123	110	96	82	69	55	41
(6) £/fisherman/year	7299	8455	9819	11 458	12 926	14 783	17 091	20 244

Notes:

(a) millions of pounds sterling.
(b) thousands.

evidence points to this being the case for the fisheries of the Community taken as a whole. For example, the arguments advanced by the industry to oppose the adoption of conservation measures are always based upon its claimed inability to sustain the short-term financial losses. The severe restrictions on making Community grants for vessel construction has led to few orders being placed for new vessels, also indicating that the industry is generating insufficient profit to cover depreciation. Perhaps most telling was the violent reaction of the catching sector of the fishing industry to its fall in incomes as a result of the slump in fish prices in early 1993.

The rate of fishing at which the maximum economic return is obtained can be determined from yield per recruit theory (Fig. 11.1) but it is instructive to attempt to quantify the sums involved. According to the 'Report 1991', the total value of landings by Community vessels in 1989 was of the order of £4300 million (ECU 6143 million). Excluding landings by Greece and making allowances for landings by France and Spain from the Mediterranean Sea as well as those from non-Community waters, the total value of landings from Community waters was of the order of £3000 million, perhaps as high as £3500 million. If the fisheries are not profitable, this is also the minimum cost of catching the fish from the waters covered by the conservation policy.

Using the generally accepted breakdown of costs as 33% for crews' wages, 60% for variable costs and 7% for depreciation, costs of servicing capital and profit, but assuming that this last set of items is not covered in the present financial state of the industry, the breakdown of income and expenditure is as in Table 11.1. Applying the results of a typical yield per recruit curve and assuming, for simplicity, no price elasticity of demand, the total income would vary as in line 1 of Table 11.1 as fishing effort decreased. Variable costs would decline in proportion to the decline in fishing effort, as line 2. Wages would vary as line 3, assuming that they remained at 33% of total income. Profit would increase as fishing effort fell, reaching a maximum when it had fallen to 60% of its present level in this example. As the number of fishermen would also decline in proportion to the reduction in fishing effort, income per fisherman would increase. This example suggests that over-fishing is dissipating potential profits of one billion pounds sterling (ECU 1.4 billion) a year. The 'Report 1991' puts it even higher at ECU 3 billion. The actual figures obtained would vary with the shape of the yield per recruit curve but the principle remains unchanged. The most important feature shown by this example is that most of the increased profit arises not from the increase in total catches but from the reduction in fishing effort, as fixed and variable costs are reduced. The principles are not new; they have been known for decades but have never even formed part of the advisory input to the management of the fisheries of the Community, let alone been acted upon.

National fisheries management

There are many, particularly in Ireland and the UK, who argue that the CFP is a fatally flawed system and that the state of the fisheries would be infinitely better if fisheries were managed nationally. Because this is seen as an objective in itself, the scenario of 'what might have been if there had been no CFP' needs to be examined.

A frequent question is why fisheries should be the only natural resource subject to a Community policy. The answer is simple. All the Member States are signatories of the Treaty and are bound by its provisions, which include the establishment of the CFP (see Chapter 2). It would be necessary for a Member State to secede from the Community in order for its fisheries not to be regulated under the CFP. However, what would have been the situation for a Member State if it had not acceded to the Community?

That there would have been fishery zones based upon 200-mile limits and median lines goes without doubt; the third UN Conference on the Law of the Sea determined that. But fish do not recognize such zones and it is common experience throughout the world that most fish stocks move between national fishery zones: they are 'shared stocks', making their joint management essential. All the most important stocks in the Community are shared stocks. It is impossible to manage shared stocks nationally because fixing objectives and determining how those objectives are to be achieved must be agreed jointly by the co-owners. Referring specifically to the UK, although these comments apply in general to all Member States, it would have been unable effectively to manage unilaterally the stocks in its waters. The UK could have laid down unilaterally the conditions such as minimum mesh sizes, by-catch limits and closed seasons, under which the fish in its waters could be caught but even this would have been of little use if other countries had laid down different measures which adversely affected British interests. More importantly, the UK could not have determined unilaterally how much could be caught from each stock, the question of TACs. The UK would have also wanted to negotiate fishing possibilities in the waters of other countries, in the same way as the Community now negotiates with the Faroe Islands, Norway and Sweden. There would have been two possibilities, either to continue negotiating within the framework of NEAFC or to engage in a series of bilateral negotiations with each country.

The experience with NEAFC showed that this was an unsatisfactory organization for effectively managing fisheries. The International Commission for the North-west Atlantic Fisheries (ICNAF) also proved ineffective in implementing fisheries conservation, as has its successor, the Northwest Atlantic Fisheries Organisation (NAFO), for the same reasons. The experience with NEAFC was that even if decisions were taken, they represented the minimalist position common to the contracting parties, and then subsequently were often not implemented because of the objection procedure. NEAFC continues to operate in this manner and has been unable to agree measures which might lead to the re-establishment of the Atlanto-Scandian stock of herring or to agree regulatory measures for blue whiting. Enforcement of the recommendations of NEAFC which were adopted by its contracting parties was entirely dependent upon national control and enforcement with no effective supervisory powers by NEAFC. At least within the Community, once a regulation is adopted, even though it may take years to achieve, its provisions are immediately binding in all Member States and the Commission can and does exercise some overall control over enforcement, limited though it may be. Continuing to operate within the framework of NEAFC would not have resulted in more effective management than under the CFP but, in fact, probably much less effective management.

The alternative scenario would have been bilateral negotiations. Because most of the fish stocks inhabit several national fishery zones during the course of their life, most of these negotiations would have had to be multilateral rather than bilateral. This might well have been the situation even for stocks inhabiting only two fishery zones. Although it might be tempting to think, for example, that Ireland and the UK could have decided bilaterally how they would exploit the Irish Sea or that France and the UK could have made a bilateral agreement concerning the English Channel, this is to ignore that each country would have made reaching a satisfactory agreement for one area conditional on maintaining its traditional fishing possibilities elsewhere. For example, France would undoubtedly have required agreements which permitted its vessels to continue to fish in the Irish Sea. This is not a hypothetical situation. The fisheries agreement between the EEC and Greenland provides for the EEC to allocate catch possibilities for deep-water prawn in Greenlandic waters to the Faroe Islands and Norway because these allocations are crucial to agreeing the annual exchange of catch possibilities between the EEC and the Faroe Islands and Norway.

Multilateral negotiations would have been less complex in one respect than the Community system because the negotiations would have been limited to those countries in whose zones the stocks lived. For example, there would not have existed the situation in which Greece and Italy, having no immediate interest in the conservation policy because it does not apply to the Mediterranean at present, can vote on conservation measures and link satisfactory solutions on other related issues, such as markets and structures, to voting for their adoption. Nor would countries in whose waters the stocks did not live have been immediately concerned except where the maintenance of traditional fishing had to be taken into account by one of the parties because it affected the outcome of bilateral negotiations concerning another zone. Apart from this, the situation would have been very similar to a meeting of the Council, but with different players, each trying to negotiate those TACs which best suited its objectives and also trying to ensure that each country adopted national technical conservation measures which were not detrimental to its own interests.

The UK would not have had a very strong hand because most of the important fish stocks in the North Sea spend a large part of their life history outside British waters. Take some examples, all concerning stocks in the North Sea. The nursery grounds of herring lie almost entirely within Danish waters which would have permitted Denmark to have a major determining role in how North Sea herring was exploited. Denmark would have always been able to threaten that, if it were not guaranteed satisfactory access to the fisheries for mature herring in UK waters, it would permit unrestricted fishing on the juveniles. The main nursery areas of cod, plaice and sole are along the western coasts of the Netherlands, Germany and Denmark, which would have applied similar pressures. Only for haddock and whiting would the UK have been in a strong position, these stocks inhabiting mainly the UK fishery zone. Given the complexities of managing the mackerel stocks under the present system, it can be only a matter of guesswork if any agreement could have been reached. It is very possible that the western stock of mackerel, which migrates between the fishery zones of Ireland, the Faroe Islands, Norway and the UK would have remained unregulated and might well have been fished out of existence.

One of the problems which the CFP has eliminated is that of policing national fishery limits. This would have been a major problem for countries such as the UK and Ireland. In particular, the CFP has enabled what would have become a difficult issue, that of the 200-mile fishery limit claimed by the UK around Rockall, to be largely ignored. It is very doubtful if the UK would have been able to maintain this claim if there had not been a CFP, let alone enforce it. Being uninhabited it does not provide the necessary legal basis under the UN Convention on the Law of the Sea for a fishery limit. At present this legal uncertainty is not an issue because it suits the majority of Member States to accept the *de facto* situation. Without the CFP, it is almost certain that the UK's claim would have been challenged in the International Court of the Hague.

But assuming that national management of exclusive fishery zones were possible, are there any grounds for believing that it would have been any more successful than the CFP? It is not as if there is no experience of national management of fisheries resources and this experience shows that the answer to this question would have been a resounding 'No'. The difficult problems of fixing objectives and taking decisions to achieve them still remain and experience shows that politicians are no better at taking these decisions at a national level than at an international level.

Those countries with almost exclusive control of their fisheries, such as Iceland and the Faroe Islands have experienced problems. One of the most heavily regulated fisheries in Europe is that for scallops in the Bay of the

Seine. This fishery is regulated by a plethora of regulations, TACs, minimum size limits, seasonal closures, periodic closures within the open seasons and daily catch limits. All these regulations have been determined by the fishermen who prosecute the fishery, yet the outcome has been disastrous. There are too many fishermen exploiting the fishery for them all to make a reasonable income so they do not respect their own rules. Within the UK, there are many stocks occurring in coastal waters, such as crabs, lobsters, brown shrimps and cockles, which are nationally managed. In every case management has been a failure, even using only the criterion of fish stock conservation.

The conclusion must be that, however unsatisfactory the conservation policy of the Community is considered, the situation would have been very little different under a nationally managed system and might have been considerably worse. Fisheries management to maximize economic benefits would have been impossible.

Summary

Fixing objectives and taking the necessary measures to achieve them is essential for effective fisheries management. Fisheries management is not simply applying the scientific advice. It is a matter of historic accident that this has become the commonly perceived view. Scientists should be required only to advise on how the objectives set by the politicians might be achieved and the biological constraints on achieving those objectives. To date, the Council has failed to set specific objectives and there appears little possibility from the latest regulation concerning the conservation policy, Regulation No. 3760/92, that it is likely to do so in the future. It seems highly probable that the fisheries of the Community will continue to be managed on the basis of maintaining an equilibrium between resources and fishing capacity, a meaningless objective, in the light of social and economic considerations, which are not defined. Additionally, Regulation No. 3760/92 provides for environmental considerations to be taken into account, which will make it even more difficult to agree upon specific objectives.

The Treaty provides specific objectives for managing the fisheries of the Community. These objectives, although phrased in terms of agriculture, can be interpreted in a manner relevant to fisheries. The objectives specified cannot be achieved simultaneously; a compromise is required. The compromise is to have an overall objective of maximizing the economic returns from the fish resources. The prize to be gained from effective management is great, of the order of one billion pounds sterling a year, possibly much more.

Chapter 12

Achieving the Objective

or Breaking the Mould

Introduction

Having fixed an objective, whatever the objective might be, the means by which it is to be achieved have to be determined. In Chapter 11 it was concluded that the objective for the CFP should be to maximize profitability. This chapter examines how this objective might be achieved within the legal constraints set by the Treaty, how the existing management mechanisms might be used and what new measures would be needed.

The CFP stems from the Treaty and its future development must respect its principles. The constraints set by these principles must be respected because they are Community law. It also has to be recognized that the CFP has reached a certain state of development; it is not possible to go back to the beginning and develop a different policy anew. What must be ignored are the political constraints which have moulded the CFP to date and which have prevented many of the actions which are required for effective fishery management from being taken. This is not to ignore that these political constraints will continue to exist and influence how the policy develops. But if they are not challenged then the policy will ossify, to its detriment.

Respecting the principles of the Treaty

The relevant principles

In revising the CFP, the principles of the Treaty must be respected. Those relevant to the policy are non-discrimination on the grounds of nationality (Article 7), freedom of movement (Article 48) and the right of establishment (Article 52).

Stemming from the principle of non-discrimination on the grounds of nationality is that of freedom of access which is provided in Article 2 of Regulation No. 101/76, in particular its paragraph 2 which states that:

'Member States shall ensure in particular equal conditions of access to and use of fishing grounds situated in the waters referred to in the preceding subparagraph (waters under the sovereignty or within the jurisdiction of each Member State) for all fishing vessels flying the flag of a Member State or registered in Community territory.'

Another principle is that of 'subsidiarity' which will become formally part of the Treaty of Rome only when, perhaps if, the Treaty of Maastricht, of which it is part, is ratified by all Member States. However, even if this treaty is not ratified, it is now politically accepted that the principle of subsidiarity should be implemented whenever practicable. The principle is that actions should be taken at the lowest, local level of government possible. In Community terms this means that the Community should not adopt legislation if the same results can be achieved by non-Community legislation without detriment to the principles and objectives of the Treaty.

Applying the principles

Relative stability: a principle contrary to the Treaty

The intended political objective of relative stability was that quotas should be caught only by fishermen of the nationality of the Member State to which the quotas were allocated. (As described in Regulations No. 170/83 and No. 3760/92, the principle is that quotas are allocated to the Member States; the regulations do not provide for who may fish them.) This principle is contrary to that of non-discrimination on the grounds of nationality and therefore incompatible with the Treaty. For this legal reason alone relative stability cannot be a principle upon which a revised CFP is based. That relative stability is contrary to the provisions on non-discrimination in the Treaty was recognized by the Court of Justice in the 'Romkes' case (see Chapter 7). In this case the Court ruled that quotas, which are the practical application of relative stability, were a justified derogation from this principle but only while the stocks were over-exploited because quotas were essential for the proper management of TACs. The Court recognized that a system of unallocated TACs results in 'a scramble for fish' and that for this reason TACs must be allocated by quotas. The decision of the Court was only justified by the circumstance that a system of TACs existed.

But there are also other reasons for abandoning this principle. Relative stability no longer operates as was intended. The application of the principles of non-discrimination on grounds of nationality, freedom of movement and the right of establishment has made it legally impossible to prevent a citizen of a Member State from establishing in any other Member State a company owning and managing fishing vessels. These vessels are entitled to fly the flag of the Member State in which the company is registered. This principle has been firmly established by the Court of Justice in a ruling against the UK in what is known as the 'Factortame' case in which UK national legislation, which was based upon nationality requirements for companies operating fishing vessels, was ruled incompatible with Community law. It also means that conditions cannot be attached to fishing licences which are discriminatory and would have the same effect.

In consequence, many vessels owned by nationals of one Member State are registered in another and their catches are counted against the quotas of the Member State in which they are registered, irrespective of the nationality of the crew members and the port in which catches are landed. For example, many Spanish-owned vessels are registered in Ireland and the UK. But the situation does not apply only to Spain; Dutch-owned vessels are registered in Belgium, Germany and the UK. Those fishermen in whose Member State the 'foreign' vessels are registered object strongly to what they consider their quotas being 'stolen' from them and the issue generates much heated political controversy. However, the situation is now an established feature of the way in which the fishing industry operates and is in conformity with the fundamental political objectives of the Community, that of the single market. The situation is no different to companies establishing factories in Member States other than their own. As relative stability can no longer achieve its intended political objective, there is no practical point in retaining it.

The accession of new Member States to the Community will inevitably create pressure to abandon the principle of relative stability. These new Member States, no more than Spain and Portugal, will not be prepared to accept a system which benefits the 11 Member States which adopted the original conservation policy and disadvantages all others. Sooner or later the principle will be successfully legally challenged, even though Spain and Portugal have been unsuccessful to date, as described in Chapter 7. The stocks will not remain overfished for ever, or should not if the policy achieves the objective of eliminating overfishing. The Community will need to implement a system which will be relevant to that situation and which copes easily with the foreseen accession of new Member States. This issue will become increasingly politically important.

Freedom of access
At present, the provisions of Regulation No. 3760/92 limit access to the 12-mile coastal zone and the Shetland Box; these provisions apply to all Member States. In addition, under the Act of Accession of Spain and Portugal, fishing by Spanish and Portuguese vessels in the waters to the west of the British Isles and in the Bay of Biscay is restricted by a system of licensing. They are also totally excluded from the area off the west coast of Ireland bounded by latitudes 50°30'N and 56°30'N, and longitude 12°W. Articles 162 and 350 of the Act of Accession of Spain and Portugal provide that these conditions of access have to be reviewed and those which will replace them adopted before 31 December 1993. The new provisions will enter into force on 1 January 1996. The restriction on access to the area west of Ireland ends on 31 December 1995; it is not renewable.

The access of all Member States to all waters is restricted in practice, if not by law, by their allocations of quotas because, if a Member State does not have sufficient quotas to make it economic for its vessels to fish in that area, they will not go there. For example, as part of the settlement of the CFP, quotas for human consumption species in the area to the west of the British Isles were deliberately not allocated to Denmark in order to limit its access to this area.

The application of the principle of 'access', in the sense in which the term is used in the context of the CFP, restricts fishing vessels of Member States to specified waters. This is discriminatory and contrary to Article 7 of the Treaty. For this reason alone the principle should be abandoned. But in terms purely of fisheries management there is no basis for maintaining it. As described in Chapter 7, access is not a conservation measure but was a response to political pressures. However, to allow unrestricted access to all Community waters by vessels of all Member States, as provided by the provisions of Article 2 of Regulation No. 101/76, would certainly result in fishing effort being concentrated in those areas in which it was most profitable to fish, leading to the over-exploitation of the stocks in these areas. Therefore access must be restricted but in a non-discriminatory manner.

Conclusions

The present conservation policy does not comply with the principles of the Treaty. A revised policy must do so. In particular, the policy is now driven by the application of the principle of relative stability. The implementation of this principle lies at the root of many of the problems which the policy experiences and prevents it being revised in order to meet what should be its objectives, a profitable industry. This principle should be abandoned both for legal and practical reasons. Politically it is realized that all Member States with the exception of Spain and Portugal are very strongly attached to this principle, as the provisions of Regulation No. 3760/92 show (see Chapter 7) but there is no point in adhering blindly to a political dogma if that dogma does not produce the practical results which everyone desires.

The methods to be used

What provisions does the CFP already provide which could be used for achieving the objective of maximizing economic benefits? How could they be used and would they be effective?

Limiting the size of the fleets

Multi-annual guidance programmes

The multi-annual guidance programmes offer an established method of reducing the size of the Community's fleets, if they can be effectively enforced. The present objective of these programmes is to bring the size of the fleets into equilibrium with resources. As far as can be determined, the target equilibrium point is that corresponding with F_{max} but there is no reason why this target should not be the rate of fishing mortality corresponding with that which would generate maximum economic benefits.

However, multi-annual guidance programmes are national programmes. They are based on the fact that each Member State is allocated a share of the resources and that the catching capacity of its fleet should be in equilibrium with that share. Their underlying basis is, therefore, that of relative stability which, as already concluded, is contrary to the principles of the Treaty. Multi-annual guidance programmes therefore cannot play a role in a CFP founded on the principles of the Treaty.

Because they are national programmes, multi-annual guidance programmes also suffer from the disadvantage that their biological basis is unsound for the reason that there is no practical model which permits the relationship between fishing capacity and the rate of fishing mortality to be determined. It is possible to determine the reduction in the rate of fishing mortality which is needed to achieve a target level of the rate of fishing mortality for any stock. It is also possible to reduce total fishing capacity until the target rate has been achieved. But allocating fishing capacity between Member States and different fleets in order to achieve the target level of the rate of fishing mortality is very difficult, if not impossible, because there is no appropriate model. This is the situation for a single stock. The problem is impossible to resolve for fisheries for a mixture of species.

The multi-annual guidance programmes are also being implemented as if F_{max} were a unique point, which it is not for the reasons described in Chapter 8. The fishing capacity appropriate to the present pattern of exploitation would not be appropriate if that pattern were changed. To take an extreme example, F_{max} for North Sea cod requires a reduction in the rate of fishing mortality of nearly 75% with the present pattern of exploitation. However, if this pattern could be improved by a massive increase in the minimum mesh size used in the fishery, no reduction in the rate of fishing would be needed.

The multi-annual guidance programmes also appear to be based upon the assumption that once fishing capacity has been reduced to its target level, the equilibrium between fishing capacity and resources will be stable. This is to ignore that, even if effectively enforced, they are an imperfect means of limiting fleet size. It is estimated that the efficiency of fishing vessels increases at the rate of 2–3% a year. This increase is achieved by improvements in design, use of more efficient gear and better equipment. As the total engine power of the fleet is limited by multi-annual guidance programmes it is not possible to increase the efficiency by increasing engine power itself. Therefore the fleets cannot be brought into equilibrium, whatever equilibrium point is chosen, and then maintained at that level in terms of tonnage and engine power, the two parameters used in the programmes. Continuous reductions in fleet size would be necessary to compensate for increases in efficiency.

Licensing

Under the provisions of Article 5 of Regulation No. 3760/92 a Community system of licensing of fishing vessels is to be established by no later than 1 January 1995. The regulation provides only that all Community fishing vessels shall be registered and that licences shall be issued and managed by Member States. Such a system means that a record of all fishing vessels will then exist. This will be a major improvement on the existing situation but licensing on its own does not resolve the problems of management. Licensing is a tool which has to be used properly.

If the provisions in the proposal by the Commission for a new regulation to replace the existing Regulation No. 2241/87 establishing a control system applicable to the common fisheries policy were adopted unchanged, the licences would have to have conditions of fishing attached to them, notably the area in which the vessel was permitted to fish. Furthermore, the Commission has proposed that all vessels having a length overall greater than 10 m would have to have equipment by which the position of a vessel could be monitored by satellite.

If these provisions are adopted, the framework for a major advance will have been established. As always though, whether the potential benefits will be realized will depend upon the effectiveness of the Member States in implementing the regulations.

Total allowable catches

TACs are a fatally flawed method for managing the mixed demersal fisheries typical of Community waters for all the reasons already described in Chapter 9.

TACs were introduced primarily to implement the principle of relative stability. As already described, this principle is both contrary to the provisions of the Treaty and is effectively inoperative. It therefore follows that there is no logical reason to maintain the system of TACs and quotas unless they have a conservation benefit. The only situation in which they almost certainly have a conservation benefit is in the fisheries for pelagic species, for two reasons. First, catches of pelagic species are largely independent of the size of the fleets. Pelagic fish swim in shoals. Once a shoal has been found the size of the catch is largely dependent upon the skill of the skipper and its maximum size determined by the hold of the vessel. In contrast, catches of demersal species are mainly dependent upon their abundance because towed gears have to sweep an area of ground while passive gears either have to sweep a volume of water, either passively in the case of gillnets, or by attracting fish in the case of long-lines and pots. Second, if TACs were not fixed for pelagic species, the fleets would concentrate upon the most valuable species, fishing them to the exclusion of the less valuable. This is possible because pelagic fish shoal by species, whereas demersal species are much more mixed.

Technical conservation measures

For the reasons given in Chapter 8, technical conservation measures by themselves not only cannot ensure that the maximum economic benefits from a fishery are achieved but, on the contrary, ensure that they are totally dissipated. The role of technical conservation measures in managing fisheries to maximize economic benefits is to ensure that these are obtained by changing the pattern of exploitation in relation to the rate of fishing mortality, as shown in Fig. 8.5 for MSY.

Conclusions

The Community has established a series of measures to manage its fisheries but has not incorporated these into a coherent system of management. This is not surprising because the Council has not agreed a set of objectives, in the absence of which effective management is impossible. The Commission has made proposals which, if adopted, will enable the conservation policy to be more effectively implemented, always supposing that the Member States show more enthusiasm for implementing these new provisions than they have the existing legislation. However, these proposals still do not constitute a coherent policy and cannot do so until there are agreed specific objectives. A total re-evaluation of the policy is needed, a re-evaluation which has the objective of effective management of the resources to achieve maximum economic benefits. Some elements of the policy need to be strengthened, others abandoned. How might this be done?

A Community system of fisheries management

How might a system which provides effective management of fisheries be implemented, one which respects the principles of the Treaty, has the objective of maximizing the economic benefits and evolves from the present system?

National or Community control: a question of subsidiarity

The majority of the fish resources in Community waters, and certainly all the most important fish resources, are not limited to the fishery zone of one Member State but are shared, often between several Member States. One of the fundamental reasons for having the CFP is the recognition of the fact that managing the harvesting of a common resource requires agreement between its owners (see Chapter 11). However, under the present system, the management of this resource is split. Management measures are adopted by the Community but their implementation is the responsibility of the Member States. This arrangement respects the principle of subsidiarity, as also will the system of licensing provided for in Regulation No. 3760/92, but is this division of responsibilities compatible with effective management?

As described in Chapter 7, the majority of Member States do not take their responsibilities seriously and, in consequence, the whole policy is failing. For management to be effective, responsibility cannot be divided. Management cannot be left to the Member States. Management must be the responsibility of the Commission; that is its institutional role (see Chapter 1). This does not conflict with the principle of subsidiarity. The principle is that measures should be taken at Community level only if the intended objectives cannot be achieved by measures taken at national or regional level.

Licensing: a means of managing the fisheries of the Community

To achieve the objective of maximizing the economic benefits from the fisheries requires that the fishing capacity of the Community's fleets be limited. A system of licensing offers the most effective means by which to achieve this objective but only if properly implemented and effectively enforced. This section develops these proposals, suggesting what is needed to achieve this objective.

Limiting the number of licences
Regulation No. 3760/92 provides for a Community system of licensing but there is no indication in the regulation that it is to be used as a means of

limiting fishing capacity. Its objective appears to be to ensure that all Community vessels are recorded. To achieve the maximum economic benefit the number of vessels must be limited, which is apparently not foreseen in the Community system. Limiting the number of licences could realize the objective which multi-annual guidance programmes are attempting to achieve. These programmes would then be no longer needed.

Licence conditions

It would be essential in the interests of efficient management of the fish stocks to attach conditions to the licence. An indispensable condition would be to specify the area of fishing for which the licence was valid otherwise all vessels would concentrate in those areas where it was most profitable to fish. This could be controlled by satellite monitoring. Both these measures have been proposed by the Commission.

It would also be necessary to attach conditions which related to the efficiency of the vessel, such as tonnage and engine power. Other conditions which might be necessary are the type of gear and mesh size but the objective would be to keep these conditions to the minimum in order to ensure that licensed vessels could fish with the minimum of restrictions. The primary objective would be to ensure that the capital represented by the fishing fleets was used as efficiently as possible. These other conditions and the need for them are described in detail later when the problem of managing multispecies demersal fisheries is discussed.

Allocating licences

The principle of non-discrimination on the grounds of nationality means that any fisherman from any Member State should have the right to hold a licence for any fishing area in Community waters, subject to the conditions described. This would meet the principle of unrestricted access as laid down in Regulation No. 101/76. It would also provide a solution to the problems both of adapting the CFP to end the discriminatory derogations on access to which Spain and Portugal are at present subject and accommodating the foreseen accession of many new Member States.

Licence fees

There is a fundamental argument for charging licence fees. The fish resources are a Community patrimony and the benefits from harvesting should therefore accrue to the Community as a whole, not just to those who are licensed to exploit them. Licence fees provide this means. This is a principle which applies to all other natural resources but is not generally recognized for fisheries for two reasons. First, fishing developed long before countries could claim ownership of the resources and therefore anyone with the means could fish them without paying; fishermen do not see why they should start paying now. Second, stemming from the first, it is widely considered that everyone has a legal right to fish the seas. In Germany this right is enshrined in the constitution, although its exact legal interpretation remains to be clarified; in theory, TACs and quotas are incompatible with this constitution.

A further argument for charging licensing fees is that the fishing industry receives a large, hidden subsidy in that the costs of managing the fisheries, such as enforcement, restructuring the industry, research and administration are a charge on the community at large. (The budget for the Fisheries Management Committee, described in Chapter 10, would come from these fees.) Doubtless the industry would not recognize this and would argue that such costs are already met by taxation. However, it remains the case that rational exploitation of the resources would generate large profits, as shown by Table 11.1, and it is only equitable that part of these should be used to meet the costs of management.

Because the profits from properly managed fisheries are potentially very large, owners of vessels would be prepared to pay for licences. Many already do in order to fish in the waters of third countries and around the Falkland Islands. There are various means by which licence fees could be set. A fee could be levied which was related to the tonnage or engine power of the vessel to which it applied, as occurs in the agreements between the Community and many third countries. The fee could also be related to the species fished because it would be more profitable to fish for some species, such as sole or *Nephrops* than whiting or industrial species. However, as there would be considerable difficulties in determining appropriate licence fees, a better system might be to offer licences by sealed tender, allowing market forces to set the level of fees. Making licences freely transferable would further permit market forces to determine their value.

There remains the question of whether the new licences should be of fixed or indefinite duration. Making them of fixed duration would mean that, on their expiry, there would be no necessity to replace them if fishing effort was still too high and there would be no costs to the licensing authority, that is the Commission. The Commission could still buy in licences earlier if it considered that fishing effort needed to be reduced more rapidly than by expiry of licences. However, as a licence would be equivalent to a property right, it would be more appropriate to issue licences of indefinite duration with the licensing authority being a buyer or seller in the market as was necessary to regulate fishing effort.

Because there would be no profits immediately the system was introduced it would not be appropriate to charge licence fees initially. However, as soon as the fisheries became profitable an increasing fee would need to be introduced in order that those holding these licences did not make exceptionally high profits.

The transitional period
While licensing can be introduced 'overnight', implementing a total system would have to be phased in over several years. This has already been recognized in discussing licence fees. The initial objective of licensing would be to limit the size of the fleet to its present level and to prevent any new entrants. It would offend against natural justice to deprive any owner, and the fishermen employed on his vessels, of their living at short notice by refusing him a licence. It would also be politically impossible to discriminate between

owners. Therefore, every vessel engaged in fishing would have to be issued with a licence initially, termed a 'transitional licence'. To permit fleet reduction without having to buy out every licence owner these licences would have to be of limited validity, the average normal life of a fishing vessel being an obvious term. It would need to be fixed in order to prevent owners maintaining their vessels in perpetuity but this might be set at say, ten years, from the date of construction. Conditions would have to be attached to the transitional licences. As TACs are ineffective in limiting fishing effort on demersal species, there is little case for retaining them for these species during the transitional period. Instead, the transitional licences might have attached to them the condition that they were valid only for the waters in which the vessel had a track record of fishing. To the extent that it was considered necessary to limit fishing effort during this period, limiting days at sea would be the most effective method but its economic inefficiency provides an argument for reducing the number of vessels by decommissioning as quickly as possible. The only stocks for which there is a case for limiting fishing effort are those of North Sea and west of Scotland cod and haddock, for which the evidence is subject to more than one interpretation (Chapter 7). An immediate benefit of allowing fishermen to land all the demersal species which they caught would be that it would eliminate one of the main reasons for discarding.

During the transitional period it would be necessary to issue new licences even though this would result in a slower reduction in fishing effort than would otherwise be achieved. Not to issue new licences would discriminate against those who owned the oldest vessels and also disrupt the shipbuilding industry. Operating a decommissioning scheme under the structural policy could be used to encourage owners to surrender these licences before they expired. The advantage of issuing new licences in this period would be that it would establish a basis for charging licence fees.

Such a system, phased in over ten years, as suggested, or even longer, would enable the fishing industry to evolve gradually from a system which is based essentially upon open access to a common resource to one of restricted access to a Community-owned resource. To the extent that the process needs to be accomplished over a shorter period than ten years in order to achieve its economic benefits as quickly as possible, decommissioning as provided for by the structural policy could be employed.

Who should manage the system?

As already argued, management of the policy must be the responsibility of the Commission. This argument applies even more forcefully to the Community system of licensing described. To permit Member States to issue licences would raise all the questions of relative stability and discrimination between nationals that the system must exclude. Furthermore, given the track record of most Member States on control and enforcement, it is difficult to imagine them implementing such a system fairly.

The Commission is well placed to run such a system. It is a little known fact that the Commission already has a unit which is responsible for issuing licences to Community vessels fishing in third country waters and the converse.

The unit is very efficient and has many years of experience which could easily be built upon to fulfil this expanded role.

Managing mixed species fisheries
To date no real attempt has been made to resolve the problem of managing the fisheries for a mixture of demersal species typical of Community waters. In theory technical conservation measures provide the solution to this problem but in practice their complexity and the inability to enforce them means that they do not.

Technical conservation measures cannot achieve optimum exploitation of all species which have different growth rates and attain dissimilar maximum sizes. Community legislation is drafted as if this were possible. One of the practical reasons for this is that there is no objective basis for resolving conflicts between fisheries for different species in which one fishery has an adverse effect on another. Decisions are determined by which group can bring the greatest political pressure to bear, each supporting its arguments with the biological evidence which best supports its case, and whose votes are required to get the adoption of compromise solutions, as described in Chapter 6. Implicit in all decision taking is that every method of fishing shall be permitted and that a fisherman shall be able to switch between one method of fishing and another during the same trip, carrying nets of as many different mesh sizes as he chooses. The Commission has proposed yet once again that fishermen shall be allowed to carry nets of only one mesh size during any one trip in its proposals for the new control and enforcement regulation.

Managing the fisheries of the Community to maximize the economic benefits would provide an objective basis for taking decisions about which fisheries should be encouraged and which should be discouraged or even stopped. Licensing would provide an effective means by which to implement the decisions by attaching the relevant conditions to the licences or not issuing licences at all for certain methods of fishing or for fishing for certain species. For example, it is possible to consider a system in which the impact of the fisheries for *Nephrops* on those for hake and whiting were limited by issuing only a restricted number of licences for this fishery. In the North Sea a proportion of the licences would have the condition that they were restricted to fishing with nets having a minimum mesh size of 140 mm or more while others would permit the use of nets having a minimum mesh size of, say, 100 mm. The former would catch only the larger species such as cod, plaice and saithe while the latter would catch most species, except sole. It is envisaged that fishermen would be able to land all fish that they caught but the balance of licences and possibly licence fees would take into account the economic damage done by those fisheries using small mesh sizes to those using large mesh sizes. The following examples illustrate the basis on which decisions between interacting fisheries might be taken.

The most long-running of these conflicts is that between those fishing for industrial species and those fishing for human consumption species. This is a case which can be examined in economic terms. Using the data in the report of the ICES Roundfish Working Group for 1990 and average recruitment for the period 1985–89, it can be calculated that banning the fishery for Norway pout would result in an additional annual increase in catch of haddock of 5000 t and of whiting of 43 000 t, with a market value of approximately £48 million, assuming no decrease in average prices as supplies increased. The value of the Norway pout catch foregone would be £4 million. Economically there is a clear case for banning this fishery. However, there is an even more pressing case for initially taking measures to eliminate the discarding of juvenile haddock and whiting, the potential additional value of the increased catch being £100 million a year.

A similar case concerns the fisheries for juvenile herring which it is argued should be banned in order to increase catches in the fisheries for adult herring. The Commission carried out an economic study of these fisheries from which it concluded that the economic benefit of banning the fisheries would be small or even negative because most if not all the increased catch would be sold for reduction to meal and oil because the market for herring for human consumption is very limited and, at present, incapable of expansion.

An example which concerns two species fished for human consumption is that of cod and *Nephrops* in the Irish Sea; cod eat *Nephrops* which has a higher unit value than cod. In this case, under certain assumptions, over-exploiting cod in order to increase catches of *Nephrops* would increase the total income from the two fisheries.

Economic studies could determine how to obtain the greatest economic return from such fisheries as those for plaice and sole, cod and haddock plus whiting and, in the Bay of Biscay, hake and *Nephrops*. In all these cases, fisheries management based upon biological considerations dictates a large minimum mesh size to catch the first named species and a small minimum mesh size to catch the second. At present the situation is resolved in an arbitrary manner, which permits capture of the smaller species. For example, the standard minimum mesh size of 100 mm for the North Sea is essentially determined by the fact that a mesh size which was much larger would retain very few haddock and whiting, but this mesh size is far too small for the optimum exploitation of cod and plaice. In other fisheries the problem is resolved by permitting derogations from the standard minimum mesh size, such as 80 mm for sole in Region 2 and 55 mm for *Nephrops* in the Bay of Biscay. In some fisheries, such as that for horse mackerel off Portugal, the smallest fish command much higher prices than the largest; exploitation of the fishery to maximize economic returns therefore conflicts with the biologically based principles of conservation. Economic assessments would provide an objective basis for deciding how the fishery should be managed while ensuring the viability of the stocks.

What is being proposed is not entirely new. A first step in managing the fisheries of the Community on this basis has already been taken with the multi-annual programmes for 1993–96 under which the fleets have been broken down by separate sectors and different reductions in fishing capacity decided for each. To implement the system described, much more detailed economic analyses would have to be available than are at present. To date, biological, not economic, considerations have determined how fisheries should be managed. As a result, there are few, if any data which would permit decisions to be taken immediately. However, once the decision was taken to manage upon this basis, the necessary research could be implemented. There would also be time to do the research during the transitional period.

The advantages of a licensing system

The political advantages
The major political advantage of licensing, that of resolving the problem posed by the continued application of the principle of relative stability in an expanded Community, has already been described. In addition, this principle commits the Community to maintaining a system of conservation based upon TACs and quotas, because this system is the only means by which the principle can be implemented, a system which is totally ineffective in managing demersal fisheries.

The principle of relative stability also causes resistance to change, leading to the ossification of the policy. With a fixed percentage of certain quotas, there is no incentive whatsoever for a Member State to consider measures which might increase total catches of all species if this resulted in an increase in the TACs of which it has a small share or no quota at all and a decrease in the sizes of those for which it has a large share. The present controversy in which Denmark advocates over-exploiting whiting in the North Sea because the multispecies model shows that it is a heavy predator on other human consumption species is a case in point. British fishermen oppose this because they fear that the gains from the increase in other species predicted by the multispecies model would not compensate for their losses of whiting, of which the UK has 52.75% of the EC share of the TAC for the North Sea.

The advantages for control and enforcement
Licensing has many advantages for control and enforcement. Vessels must enter port where it is physically very easy to check licences; it is notable that, in the report of the Commission on enforcement, control of logbooks is the most frequent inspection. Unlicensed vessels could be seized immediately. Control of the type of gear is also easy although the amount much less so. The area of fishing can also be monitored by the use of transponders interrogated by satellites, as proposed by the Commission. The presence of a vessel in an area for which it did not hold a licence to fish or through which it did not need to steam to an area for which it held a licence would be *prima facie* evidence of an offence. Suitable algorithms could be written to determine whether a vessel was steaming through an area for which it did not have a licence.

The automatic withdrawal or suspension of a licence could be made the main penalty for contravention of regulations. Suspension of a licence for which a substantial fee had been paid would constitute a very effective deterrent and would minimize the need for expending so many resources at sea. The period of suspension could correspond to the gravity of the offence, with permanent withdrawal for repeated offences. Furthermore, if penalties were fixed by Community legislation, it would remove what is at present a serious anomaly of what is supposed to be a Common Fisheries Policy, that fishermen who commit offences against Community legislation are often subject to very differing fines in national courts for the same offence. This is a case of discrimination which fishermen find unjust and a ready excuse for not respecting the regulations. The Commission has notified its intention to make proposals along these lines if its proposals on licensing in the replacement control regulation are adopted.

Economic difficulties are probably responsible for much of the cheating which presently occurs. Limited licensing would make the fisheries profitable and reduce the incentive to cheat, although there will always be fishermen who will try to make more profits than can be obtained legally. One of the main reasons for which Community legislation, especially technical conservation measures, is so complex is that they have been framed in order to make the smallest economic impact on the activities of fishermen. This is the reason for which there are so many derogations. Making fisheries profitable would remove this economic argument and permit much more stringent but simple regulations to be adopted. That those fishermen who hold a licence will benefit from any measures taken provides an incentive to respect them; at present, any benefits arising encourage more fishermen to enter the fisheries or existing fishermen to increase their efforts.

The disadvantages of a licensing system

The lack of a scientific model
A major problem in instituting a system of limited licensing is that there is no scientific model by which rates of fishing mortality can be related to fishing capacity. Reductions in capacity would have to be made and their effects monitored. This is discussed subsequently in the section 'The role of fishery science'.

The lack of such a model has not prevented the Community from adopting multi-annual guidance programmes. It might also be said that it has not stopped the Community from using a system of TACs under circumstances in which it is not appropriate to do so.

Problems with third countries
Implementing a system of licensing would be far from problem-free, the main problem specific to the CFP being that the sharing of the resources for jointly owned stocks is now done on the basis of quotas. If Norway and Sweden were to accede to the Community the problem would resolve itself. However, although Sweden has made an application to accede, which the Commission has approved, Norway remains hesitant. However, even if Norway does not accede it seems very probable that, in the framework of the European Economic Area, it would be prepared to switch to a licensing system.

Entrenched attitudes
A major problem would be that of changing the attitude of much of the catching sector of the fishing industry which is still largely determined by the belief

that the fish resources belong to them, that they have an almost unfettered right to fish them and that, if problems arise, these are the fault of governments or the Community which should provide subsidies in order to compensate for the latter's mismanagement. There is also a strong belief that sons should be able to follow their fathers into fishing as of right and that coastal communities whose existence depends upon fishing must be preserved at whatever cost. The latter view is shared by politicians because they would foresee a high political price to be paid for changing the system. As demonstrated by the attitude to relative stability, the fish resources are considered a national resource which should be fished by fishermen having the nationality of the Member State to which the quotas are allocated in order that the financial benefits should accrue to them. However, the system of TACs and quotas does not guarantee that all the economic benefits accrue to the fishing industry of the Member State to which the quotas are allocated.

Much of this attitude that the fish stocks are a national resource which must be fished by fishermen of the coastal state and landed in the coastal state does not correspond to reality. Vessels owned and entirely crewed by nationals of one Member State frequently and often continuously land in another Member State in order to profit from the higher prices obtainable, ignoring the economic consequences for the national processing industry. Landings are often made in one Member State and freighted to another. All these moves are dictated by the economic advantages which can be gained.

While there are many fishermen in the Community who undoubtedly fish only in their national fishery zones, there are many who are equally dependent upon fishing in the fishery zones of other Member States. This is particularly so for the fisheries for pelagic species for which fishermen must be able to follow the migrating shoals.

Protecting coastal communities
Almost certainly the most sensitive issue is that of coastal communities which are dependent upon fishing. It would be strongly argued that under a system of licensing, the licences would be acquired by large companies to the detriment of individually owned boats. It is also contended that this would inevitably result in the demise of many small coastal communities, for which fishing is the main, often the only, means of livelihood. The Economic and Social Committee, commenting on the 'Report 1991' stated that licences should not be made transferable for this reason. Clearly there are economic costs which result from unemployment. Although the objective specified in Chapter 11 was essentially that of MEY, the actual objective specified was that of maximizing economic benefits, an objective deliberately chosen in order that account could be taken of costs, such as social costs, in developing the policy towards this objective.

Whether fishing should be done by individually or company-owned vessels is an emotional issue but as it is probably easier to find or create alternative employment in large ports, where companies are likely to be based, than in small ports, it might be appropriate to reserve a proportion of the licences for boats based in the latter areas. This would be a matter for political decision and would not be contrary to the principle of non-discrimination on the grounds of nationality. In fact, many of these communities would be best placed to take advantage of the proposed system because they are closest to the fishing grounds and have the lowest operating costs. These communities should also benefit by becoming the main centres for landing and processing fish as well as shipbuilding and repairs. It does not make economic sense to transport fish long distances by sea in the ships which are designed to catch them. It would be far better to land catches as close as possible to the point of capture and transport them to the point of processing. It would be better still to process the fish at the point of landing and export added-value products. This development is already occurring. French vessels fishing in the deep waters to the west of Scotland are landing their catches at Lochinver on the west coast of Scotland, although the catches are still transported to France by lorry.

But not every community can or should be maintained. To argue that they should is to argue, for example, that the Common Agricultural Policy should have been operated to maintain every inefficient farmer that existed at its inception. Farming has changed out of all recognition since 1957 with a massive reduction in the number of persons employed and the disappearance of many trades which were dependent upon horses as its main source of power. Other Community industries, notably coal and steel, have also had to be drastically restructured; there is no reason why the fishing industry should be a special case.

As a final note, it is worth remarking that what is proposed has largely come to pass for the fisheries for pelagic species. These fisheries were prosecuted for centuries by a large number of small fishing boats using driftnets. The collapse of the herring stocks brought about their demise. Now the fisheries are prosecuted by only a small number of purse seiners and pelagic trawlers which are already licensed and work within the constraints of quotas.

The role of fishery science

The system would have to have a biological basis. However, instead of it operating as at present with the fishery scientists producing reports on which the Commission and the Council acted, by adopting measures to achieve certain biological objectives, the Council would set the objectives. The fishery scientists would then advise on how those objectives could best be achieved and the biological limitations upon achieving them. The Council, acting on a proposal from the Commission, would then adopt appropriate measures and the fishery scientists would monitor the stocks to determine the effect on the stocks and whether the objectives were being achieved.

To illustrate with an example how the system might work, the STECF might advise the Commission that the economic returns from the fisheries for North Sea roundfish could be maximized by using an appropriate mixture of minimum mesh sizes and varying reductions in rates of fishing. Using the consultation system described in Chapter 10 the Commission might then propose that a certain number of licences, each having as a condition of the licence a specified minimum mesh size, be issued. This would be done and the fishery scientists would then monitor the stocks to see what happened. In the light of their reports the Commission would act, possibly by issuing more licences in some cases and buying in others. Once the initial major reduction in fishing rates had been achieved, it would be a matter of fine tuning which could be achieved with minimum disruption to the activities of the industry. The frenetic period during which TACs were fixed would become a thing of the past. TACs would still have to be fixed for pelagic species but the advice for these species is usually available by May. Implementing the new system would involve applying economic data to the existing biological models.

The future of other policies

That the conservation, markets and structural policies must be operated as an integrated set of policies is recognized in Regulation No. 3760/92. However, setting the maximization of the economic benefits as the objective of the conservation would have far-reaching consequences for these other policies.

Structural policy

The structural policy would play a major role in restructuring the industry by providing decommissioning grants to encourage owners to surrender licences. These grants would need to be set, not on the arbitrary basis on which they are now fixed, but on carefully researched levels which would provide a strong economic incentive to owners to surrender transitional licences. At the same time the Social Fund would need to be used to ameliorate the social problems which would arise.

Once the fleet had been restructured and the objective of maximum economic benefits had been achieved, there would be no need for a structural fund which is only a means of subsidizing an uneconomic industry. A highly profitable industry would be self-financing. Although this book does not concern aquaculture, the same philosophy should apply to that industry.

Markets policy

The same comments apply to the markets policy as to the structural policy. This policy is also a means of subsidizing the industry. The decline in the size of the budget for this policy demonstrates its already decreasing importance and it would become totally unnecessary in the circumstances foreseen. Many of the existing support prices would be already unnecessary if the catching side of the industry rationalized its activities instead of continuing with a system in which landings were made irrespective of demand with the oversupply of the market which so frequently occurs. The industry should finance this itself.

External resources

The policy for external resources does not affect that for internal resources except that imports of fish at low or nil tariffs will reduce the prices of fish caught by Community vessels. However, it is improbable that, even if the full benefits of fisheries management were obtained, the deficit in internal supplies would be ended. Supplies from third countries will still be required in order to ensure meeting the objective of 'reasonable prices for consumers'. As already recognized by the Commission, these supplies are likely to derive more and more from imports, probably originating from joint venture agreements, because third countries will no longer be prepared to continue to enter into agreements for Community vessels to fish in their waters nor the Community to pay the price demanded. It would be expected that this policy would be retained, limited only by how much the Community is prepared to pay to fish in the waters of third countries and by the willingness of third countries to enter into fisheries agreements.

Breaking the mould

To change to the system described would be to break the mould which has until now formed the basis for what is considered to be 'fisheries management'. To do so would require a fundamental change in attitude by fishermen, politicians and administrators. Most importantly, it would require recognition of the fact that the principle of relative stability, which is now considered the fundamental, untouchable keystone of the conservation policy, is in fact at the heart of all the problems which confront the conservation policy. Despite the fact that this principle is not achieving its intended objective and despite the enormous consequential problems which stem from its maintenance, there seems no possibility of it being abandoned. The 'Report 1991' argued strongly for its retention and, as described in Chapter 7, the precise wording in which the principle was described in Regulation No. 170/83 has been retained in Regulation No. 3760/92. With this principle firmly entrenched until the end of 2002, it seems inevitable that the CFP will continue to evolve along the same lines as it has done to date. Even if the proposals made by the Commission in its proposed new control and enforcement regulation are adopted, they will achieve little.

The prospects will be bleak for most fishermen in the Community. Within the next decade the Common Fisheries Policy must be reformed if it is to survive as an acceptable policy. The portents are not good and time is short.

Chapter 13

Summary

or Drawing the Threads Together

The CFP suffers from four major defects. First, it was not conceived as a coherent, integrated policy. Consequently, the policy has always contained a major internal contradiction, that between the structural and conservation policies. Second, the policy has no specific objectives. It has never been possible, therefore, to implement it effectively. Third, the fundamental basis for the policy has been biological, not economic. Fourth, the policy suffers from the decision-taking processes of the Community under which every decision is a political compromise (Chapter 1).

That there are no specific objectives has meant that decisions have been taken in the light of the political imperatives at the time. The markets and structures policies were adopted in 1970 in order to resolve the problems created in France and Italy by the operation of the Common Customs Tariff (Chapter 2). The conservation policy, although conceived as a policy to ensure the effective management of the stocks, rapidly became a means of responding to the political problems of access and sharing of the resources, the problem of relative stability.

The first steps in developing the CFP, taken in 1970, were to adopt the structures and markets policies. The markets policy has rarely proved controversial because it has had little negative impact on the industry. It has also been relatively successful. There have been two reasons for this. First, the Community has a huge deficit in supplies of fish, imports being 37 per cent by weight of the tonnage of landings in the Community. The value of imports is equal to that of landings (ECU 6.4 million in 1990). The consequence has been that fish prices have remained high, except for occasional short periods of oversupply. Second, the policy has always been operated to discourage fishing for intervention. This policy is essentially independent of the other three.

The main thrust of the structural policy from 1970 until as late as 1991 was to construct new vessels and modernize old ones (Chapter 2). This might have been a rational policy up until the extension of fishery limits to 200 miles. Community vessels had to compete with the fleets of other countries in international waters and, to paraphrase a famous motto, 'He who fished hardest, won'. But from 1977 onwards there was no excuse for continuing to operate the policy in this manner. After 1977 the Community had almost total control over the stocks in its waters and was in a position to manage them rationally. It was known that the fish stocks were heavily over-exploited and that less fish, not more, would be obtained by fishing harder. However, the policy put money into the pockets of fishermen and provided employment for boat builders. It was, therefore, politically popular and there were no problems in continuing its operation. Even when the Community started to develop its conservation policy and the inherent contradiction between the two policies became very evident, it was the case for the structural policy which politically won the day. The conservation policy was politically very unpopular. It was considered that it would prevent fishermen from catching fish, resulting in them earning less money, at least initially, and might put some out of work. This was not politically acceptable.

When the Community eventually started to develop its conservation policy, it was not the issue of efficient management of the fish stocks which was at the forefront of ministers' minds but the political issue of sharing the resources (Chapter 3). This problem was created by a combination of factors. The provisions on access which existed when Denmark, Ireland and the UK acceded to the Community, combined with the extension of fishery limits to 200 miles, created a Community sea to which the fishermen of all Member States had unrestricted access, except within the 12-mile coastal zone. The fishermen of Ireland and, more particularly, the UK considered that this had robbed them of the benefits which they had expected to obtain from the extension of fishery limits. These fishermen regarded the fish resources within the extended limits as a national heritage, even though, up until the extension of the limits, the waters had been international and the fish therein belonged to no-one. The debate about the conservation policy was never a debate about how to manage the fishery resources efficiently, but how to find a solution to this political problem.

Access had nothing to do with fish stock conservation even though the Shetland Box was introduced on the basis that it was to protect 'species of special importance in that region which are biologically sensitive by reason of their exploitation characteristics'. The Shetland Box was simply part of the political price paid to the UK in order to get the conservation policy adopted. TACs were nominally a conservation measure and were presented as such but, in fact, formed a convenient solution to the political problem of relative stability. The Member States wanted to share the resources between them. They needed a means upon which they could agree in order to do this. Tonnages were a measure common to all which could be understood by all. Fortuitously, fishery scientists had just developed the means by which catch possibilities could be calculated, upon which TACs are based. TACs had already been in use for a few years by the international fisheries commissions so ministers and administrators were familiar with them. TACs therefore provided the perfect opportune tool to use in resolving a political problem. They also had the added advantage that their adoption appeared to be a conservation measure. This ignored the fact that their conservation benefit depended upon how they were implemented.

Technical conservation measures were the third part of the conservation package (Chapter 3). For the most part, the Community simply adopted those measures which had been in place under the old NEAFC regime. Adopting these old regulations was relatively uncontroversial and enabled the conservation policy to be agreed. The measures were uncontroversial because they were almost unenforceable, so adopting them caused no problems for the fishermen and consequently no adverse political feedback.

Control and enforcement measures were the fourth element of the conservation policy but it was only the UK which had a real interest in this issue (Chapter 3). Although measures were agreed, they were essentially procedural measures with no means by which they could be enforced by the Commission. The Member States retained responsibility for implementing the regulation. As events were to show, this meant that the provisions of the regulation were largely not implemented.

Despite all the problems with the conservation policy, it has been a political success (Chapter 7). Even in practical terms, the policy cannot be described as a disaster. It is true that the stocks have not been rebuilt but it is not the operation of the policy which has resulted in the decline of the stocks; that has been a consequence of natural falls in recruitment over which man has no control. Catches from many stocks have increased, but again this is to be credited to increases in recruitment, not to the operation of the policy. In practical terms, catches of demersal species might not have been much different if there had not been a conservation policy. On the other hand, the policy has almost certainly limited the exploitation of the pelagic stocks.

The only policy which can be described as an almost unqualified success is that for international fishery relations (Chapter 2), essentially because it did have a clear objective, to allow Community vessels to continue fishing in the waters of third countries. This met the requirements to maintain supplies of fish from these waters and to reduce the Community's deficit of imported fish and to maintain employment in the distant water fleets. This policy had the added advantage that it was independent of the other policies, apart from having the objective of ensuring that vessels fishing in third country waters did not return to Community waters, thus increasing fishing pressure on the stocks therein. Politically, the Community was prepared to pay the price of these agreements. Perhaps the only query against this policy is that the Community paid too much.

There was a good political reason for adopting the anodyne set of measures which comprised the conservation policy – it enabled the CFP to be completed. The situation might have been chaotic if it had not been completed prior to the end of 1982 because the provisions of Article 103 of the Act of Accession of Denmark, Ireland, Norway and the UK expired on that date and needed to be replaced. As it was, there was a turbulent period of 25 days created by the fact that the conservation policy was not adopted until 25 January 1983 (Chapter 3). There continued to be good political reasons until the late 1980s for not attempting any radical revision of the policy (Chapter 4). Not that any Member State wanted to do this. While catches remained high and fishermen were generally satisfied, it was politically more expedient to leave matters as they stood. Although the majority of stocks were over-exploited, the high level of catches from the most important stocks did not make it evident that anything was amiss, despite the scientific advice that rates of fishing on most stocks were far too high. Nature conspired against any reappraisal by providing exceptionally large year-classes for many important stocks for several years

(Chapter 7). When nature finally stopped providing its bounty, the true state of the stocks became rapidly apparent and the Community was jolted into taking action. Even then it did the minimum possible (Chapters 4 to 6).

The decline in the stocks eventually forced a reappraisal of the structural policy. It at last became accepted that the structural and the conservation policies were totally interrelated and that the way in which the former was being implemented was to the utter detriment of the latter. Multi-annual guidance programmes were introduced which had the objective of reducing the size of the fleets, although the first stringent measures were not taken until the adoption of the programmes for the period 1993–96. These programmes were backed up by a marked reduction in the number of projects for construction of new vessels which were financed (Chapter 2).

However, the Community has still not come to grasps with the problem of how fisheries should be managed. Fisheries management is still seen primarily in terms of fish stock conservation, that is, taking measures whose objective is to increase the size of the stocks, without specifically taking account of the consequences for the fishing industry. Regulation No. 3760/92 does recognize the need to integrate the markets, structural and conservation policies and, in particular, is much more specific than Regulation No. 170/83 on the necessity to limit fishing capacity, but this is primarily seen in terms of its effects on the stocks. As argued in this book, fisheries should be managed on the basis of setting specific objectives for the fishing industry and then taking the measures which would permit these objectives to be realized (Chapter 8). The effect of fishing on the stocks has to be taken into account in setting and achieving the objectives but fish stock conservation is one of the tools of management, a means to an end, not an end in itself.

That fisheries management is still seen in terms of fish stock conservation is largely a matter of historic accident. Before the extension of fishery limits to 200 miles, fisheries management could be based only upon achieving biological objectives, essentially F_{max}. The advice given by fishery scientists was dominant, although often not acted upon. It was dominant because it was well researched, well presented and was virtually unanimous. This situation did not change with the extension of fishery limits or with the adoption of the conservation policy. The scientific advice provided the cornerstone of the policy, although there were always socio-economic reasons for not implementing it. To have done so, it was argued, would have reduced fishermen's incomes. The worse things became, the more drastic became the measures which needed to be taken and the greater the economic arguments for not taking them. This explains the long delays in getting conservation measures adopted and the reason why most of those adopted could never be implemented effectively (Chapter 6).

At the start of 1993 the CFP appears to be at a crossroads. On the one hand, the adoption of Regulation No. 3760/92 would appear to indicate that the policy is to continue along the same road as it has followed since 1983. In particular, the principle of relative stability has been retained totally unchanged. The concomitant of that is the retention of the system of TACs and quotas, despite the problems associated with operating this system and its ineffectiveness for managing fisheries for a mixture of demersal species. Regulation No. 3760/92 provides for multi-annual and multispecies TACs as a means of resolving this problem, although their probable primary objective is to resolve the political problem caused by having to stop fishing when a quota is exhausted. These measures will almost certainly make a bad situation worse (Chapter 9). On the other hand, the adoption of rigorous multi-annual guidance programmes, the emphasis of the structural programme on decommissioning and the introduction of licensing by 1995 appear to indicate that the policy is to follow a new road, in many respects similar to that described in Chapter 12. Additionally, surveillance by satellite has been proposed by the Commission. However, there still remain two major problems. First, there are no clear, specific objectives, let alone agreed objectives. Second, the responsibility for management remains divided; the Community makes the rules but the Member States retain responsibility for enforcing them, as well as for running the licensing system. Experience has shown that this is a recipe for disaster. For management to be effective it cannot be divided. It is argued that the fisheries of the Community must be managed by the Commission. This is its institutional role (Chapter 12). Thus it is very uncertain which route the policy will follow. The lack of objectives also means that no thought has apparently been given as to what would be the consequences of following the new route. If the multi-annual guidance programmes are successful then there will be fewer, but richer fishermen. If these fishermen become too rich, there will be increasing political pressure for more vessels to be constructed with the likelihood of the whole vicious circle being restarted. That the fish resources of the Community should be considered as a common patrimony, whose exploitation should benefit all, not just a select group of fishermen, does not even appear to have been considered.

Regulation No. 3760/92 provides many vague objectives, all admirable but all couched in such general terms as to mean all things to all men. And yet there is one objective which all fishermen want, which is to have a reasonable standard of living from operating in a profitable industry. If the fisheries were managed on this basis there would be continuous, relatively stable supplies, to the benefit of processors while consumers should also benefit from reasonable prices. It is argued in this book that this should be the objective and that it is, in fact, the objective for the CFP set by the Treaty (Chapter 11).

If this specific objective were adopted and its implications analysed, the policy could be developed to realize it. Fishery science shows the means by which it can be realized, which is primarily to reduce the amount of fishing by limiting the number of fishing vessels. This should be achieved by a system of limited licensing. The licensing system must be operated by the Commission and licence fees charged in order to cover all the costs of managing the fisheries as well as to ensure that a relatively small number of fishermen do not gain all the benefits of effective management. These benefits are large (Chapter 11). If the stocks were well managed they could yield profits which are estimated to be one billion pounds sterling a year, or even three times this according to the Commission. The capital value of the Community's fleets not being available, it is not possible to state how much of this profit would represent a reasonable return on capital, but even if this represented half, £500 million would be available for all the costs of administration, control and enforcement and research. It would cost money to achieve this objective. How much is difficult to calculate but an estimate can be made. The UK industry reckons that £150 million in decommissioning grants would be needed to reduce the UK fleet by 20%, the reduction required to meet the objective of the UK's multi-annual guidance programme. The size of the UK fleet on 1 January 1992 was almost one seventh of the total Community, excluding the Azores and Madeira, measured in kilowatts. Therefore, an approximate cost of reducing the Community's fleet to the objective set by the multi-annual guidance programmes would be of the order of one billion pounds sterling or ECU 1.5 billion. Reducing the fleets to the size needed to achieve MEY might require a reduction of twice this amount (see Table 11.1). In terms of the Community budget this is minuscule. For example, the cost of the Common Agricultural Policy is ECU 3,600 million a year (1992 budget) annually; the cost of decommissioning, on the other hand, would be a one-off sum. The sum would not be required in one year but could be spread over, say, five to ten years; decommissioning would need to be gradual. The benefits would not be immediate but would take time to build up. This period would be related to the growth of the stocks but within five years substantial benefits should be apparent. Admittedly all these calculations are approximate but the two figures of potential benefits has been calculated independently and are of the same order. The costs of achieving the policy might be greater.

What is being proposed is not novel in terms of Community policy. The Community has spent vast sums on restructuring the coal and steel industries. The Community also spends enormous sums each year on running the Common Agricultural Policy, in many instances to produce food which is surplus to the needs of the Community and which, sold on the world markets, depresses prices and causes immense political problems. In contrast, restructuring the fishing industry would result in almost nothing but benefits. Fish is an infinitely renewable resource which costs nothing to produce. The Community has a deficit of fish supplies so it can utilize all the fish which it catches. Once what would be a relatively insignificant sum in terms of what has been spent by the Community on restructuring the coal and steel industries and what is spent annually on the CAP has been expended, no further payments would be required. The policy would become self-financing. There is a political price to pay. Some fishermen would lose their livelihood and unemployment might rise to unacceptable levels in small, isolated communities. The former can be resolved by restructuring the industry gradually, allowing fishermen to leave as they reach retirement age. It would be essential to allow some new entrants, otherwise the age structure of the fishermen would become imbalanced. Expenditure might well be needed to assist the isolated communities, although as argued in Chapter 12, most of these are those well placed to gain from the proposed policy. However, if finance is required, licence fees would provide the budget.

It is ironic that ministers who are much concerned with public expenditure appear to be still trapped in an historic time warp and do not realize that, by making what would be a relatively small outlay in terms of the Community budget, the fishing industry could be transformed into a flourishing industry which would no longer be a drain on the public purse but would contribute to it. Fishermen would then have the 'fair' income which the policy should provide them and processors a high level of supplies which were stable, as provided by the objectives set by the Treaty. That would be a Common Fisheries Policy worth having.

Chapter 14

The Common Fisheries Policy – Now

David Garrod

Recent events

People have at last become aware that the EU fish stocks have been chronically overfished for many years, but the realization may have come too late. There is a tendency to blame this parlous situation on the shortcomings of the CFP and especially the conservation policy, but, as Holden makes clear, the CFP does not stand in isolation. It represents the culmination of at least 25 years of fishery management within NEAFC and ICNAF, and, as with the Icelandic cod dispute, in association with them. The framework was already well established even before the run-up to the CFP in 1983 and had been negotiated over a number of years by the same parties, if not the same people. On the downside, that also meant the CFP was surrounded by negotiating positions that were already well established. It was hardly the time for new thinking, and in any case the people concerned saw no reason for it. Everyone involved, including industry, agreed it was a sound basis in the light of experience and knowledge at the time, and such wisdom as there is now has only come with the benefit of hindsight.

Three key elements of the CFP that were new, the structures, markets and external policies, have remained largely unchanged since 1983, except for some gradual evolution. But the conservation regime has remained a source of controversy. There are problems inherent in the annual negotiation of catch allocations and adjustments to the technical measures, and there have been particular complications in the adaptation of the *acquis communitaire* to facilitate the integration of Spain and Portugal, and the further enlargement of the EU. Specific issues and policy developments to build the CFP into a more effective body of legislation have been as follows:

(1)   The mid-term review 1992

The mid-term review was scheduled to take stock of the position at the 1983–2002 mid-point. It initiated far reaching discussion on:

(a)   the development of a four-year management strategy to set management objectives for individual stocks, limiting entry and introducing a licensing scheme as a basis for the direct control of fishing effort and to elaborate arrangements for the following points (b), (c) and (d);
(b)   more comprehensive monitoring of fish catches and fishing activity, including satellite surveillance;

(c) multi-annual and multi-species TACs; and
(d) explicit consideration of socio-economic factors within management deci-
     sions.

Holden has already summarized the main points of the mid-term review
(Council Regulation (EEC) No. 3760/92); he saw these as a statement of intent
towards his own vision of effective conservation, but three years later there are
as yet no agreed proposals for any part of it.

(2) *The Multi-Annual Guidance Programme (MAGP)*

The MAGP I–III 1991–1996 were intended to stabilize and then reduce fishing
capacity by specified targets and were then to be followed by MAGP IV (1996–
1999) to bring capacity into phase with EU structural fund programmes. Member
States have had the option to reach their target reductions by decommissioning
plus direct control of effort at the national level if they saw fit. The EU claims the
programme is on target overall, but, as Holden points out, the implementation
has been uneven. No-one has yet made use of the provision for direct effort
control – the UK tried to limit days at sea but failed – and the varied success of
decommissioning and the types of vessels selected does lead to suspicions of
creative accounting. At present it seems doubtful that the 1996 outcome will
fully meet the target reduction of 8% and this figure was already only a very
feeble compromise on what scientists have advised as being necessary in conservation terms. There is no evidence that capacity reductions are having *any*
effect on real effort and fishing rates.

Fishing capacity is denominated as kilowatt hours and thereby provides very
wide scope for manipulation. These doubts on equitable measurement of fishing
power and its calibration in terms of effective fishing effort have always been the
major objection to direct effort control. Some of the practical difficulties sur-
faced in the UK industry's 1994 rejection of national proposals to limit days at
sea in support of the UK TAC allocation.

(3) *Spanish and Portuguese integration, and enlargement*

Further evidence of practical objections to direct effort control was also revealed
in the negotiation for the full integration of Spain and Portugal into the CFP to
date from 1 January 1996. In effect, Spain sought equivalent terms with Portugal
for full integration before the EU might be enlarged.

The adaptation of the *acquis communitaire* more or less maintained the balance of national TAC allocations, but the conditions for increased access to EU
waters and to the 'Irish Box' proved as controversial as everyone expected. It was
agreed that forty Spanish vessels could have access to ICES areas VIIg, VIIj, VIIb,
south and west of Ireland, and VIa, west of Scotland within the original 'Irish
Box', but would be excluded from VIIa, the Irish Sea, and VIIf, the Bristol
Channel and enforcement would be dependent on 'hail and catch reporting'
arrangements rather than more specific effort control.

The Spanish and Portuguese accession was but one example of a continuing series of disputes over the rights of access to fishing grounds. So far as the EU is concerned, though the fisheries aspects of the accession of Sweden and Finland presented less difficulty, the formula for Norwegian accession was characterized by complex problems over rights of access to mackerel fisheries west of the British Isles and EU reciprocal access to fisheries off the north of Norway and at Svalbard, and the final authority on management measures in what, after accession, would have become EU Arctic waters. The decision by Norway *not* to join the EU might prove to have been the most important single event to affect the CFP in recent years, influenced, as it was, by perceived shortcomings in ownership, access, enforcement, etc. – in effect, the poor showing of the CFP for the purpose intended, especially conservation.

There are continuing disputes over access to a cod fishery in international waters in the Barents Sea, to Atlanto-Scandian herring which has extended its migration routes as the stock has recovered, and over tuna in and beyond exclusive fishery limits west of the Bay of Biscay. The most acrimonious dispute arose between Canada and the EU, acting on behalf of Spain, concerning access to Greenland halibut stocks straddling the international boundary on the Newfoundland Grand Bank, and the enforcement of agreed regulations.

(4) TACs and technical measures

The developments described above have taken place against the annual cycle of negotiations on catch quotas and technical measures. Technological improvements, e.g. twin trawl rigs and bridge electronics, have increased efficiency (i.e. killing power), but the design of conservation friendly gear, e.g. separator trawls and square mesh cod-ends, has not been received with unqualified enthusiasm. The main concern is, I suspect, that this aspect of improved technology might actually be successful, resulting in a short-term loss of catch (for effective conservation), but a less acceptable reduction in earnings. It is, of course, too easy to criticize the industry for being unwilling to accept a short-term loss for a longer term benefit because all of us would react in much the same way. But whilst the net outcome of the lack of change has maintained the status quo for fishermen, it has been detrimental for the stocks. Herring and mackerel have begun to decline again; cod and haddock are just holding their already very reduced levels; the level of fishing on Dover sole is close to historic high levels, with the North Sea stock falling back as recent good year classes pass out of the fishery; and plaice are becoming more seriously overexploited, necessitating a reduction in TACs.

(5) Non-fishery related aspects of marine ecosystems

Other developments have taken place which also have important implications for the CFP, even though they have not yet had a direct impact. These include the Fourth Ministerial Conference on the Protection of the North Sea held under the aegis of government departments with environmental rather than fisheries responsibilities. Among other things, this echoed fishery departments' increasing recognition of the importance of interactions between fisheries and fish stocks themselves, and interactions between them and other marine resources which should eventually be drawn together in 'ecosystem management'. This concept is not yet very clear in scientific and practical terms, but it was incorporated in principle at least in the Convention on Antarctic Living Marine Resources. Respected NGOs, including the RSPB, have certainly expressed concern over the effect of fisheries on sea birds and marine mammals. The North Sea conferences have also reviewed marine pollution and concluded that fishing itself is by far the most deleterious anthropogenic influence on fish stocks and other marine resources. Particular attention has been drawn to the effect of fishing on the benthic fauna and its implications for biodiversity and the potential need for closed areas to protect complete ecosystems. Altogether these place a greater urgency on existing fisheries management to be and be seen to be in effective control of exploitation of living marine resources.

Other related studies include the UK Panel on Sustainable Development which floated the idea of an Inter-governmental Panel on the Oceans, and, in the UK, the House of Lords Select Committee on Science and Technology Report on the Fishing Industry. None has yet provided light at the end of the tunnel.

(6) The overall position

While it is true that the recent agreements with Spain and Portugal have been politically successful in maintaining the access and 'relative stability' principles within the CFP conservation policy, yield has only been maintained as a result of fortuitous recruitment of better year classes. The level of exploitation is not under control and, as always, there is no obvious will on the part of managers to do more than maintain a status quo that can be accepted by industry. The resources themselves are being eroded, causing increasing variability and decreasing economic efficiency. My own fishmonger can no longer offer the quantity or variety of European fish that was standard only a very few years ago, and this says much more than the official or even scientific statistics.

It follows that I can add nothing to the conclusions and overall assessment set out by Holden on page 167. His view on the performance of the CFP remains absolutely correct and this is despite the now accepted certainty that an effective conservation policy is the *sine qua non* of economically sustainable fish production.

Enforcement, ownership and structures – the reality?

It is possible to discern what I would call strategic principles as well as tactical arrangements within the CFP. The tactical arrangements concern the specific regulations on access, TACs, technical measures, etc. The strategic principles determine ownership and access to resources, the basis for enforcement, and links between structures and investment in fishing capacity. The situation described in the previous section concludes that the CFP is not working, and while it is customary to blame the tactical arrangements, it is worthwhile to revisit the strategic principles to see if they are adequate for their purpose, because together they provide the 'environment' for conservation in fisheries management in the EU and indeed elsewhere in the world. There is also, now, an increasingly important fourth element – the credibility of the system for both the manager and the managed.

Enforcement

Conservation regulations are inevitably restrictive, and the policy can only be successful if there is effective enforcement. The orderly conduct of affairs depends on the willingness of people to comply with the rules and most will do so if it is in their own interests. Those interests in turn depend on the chance of an infringement being detected and the penalties that go with it. In the EU the centralized 'European' ownership of resources is reflected in partially delegated enforcement where at both national and international level the inspectorates are totally inadequate for the task required of them. The inadequacies may, of course, be deliberate policy because, for some, more rigorous enforcement could aggravate grievances within the fishing community. Indeed, there is a widespread sympathy for hard-pressed fishermen and it is not unknown for authorities to disregard blatant infringements. This set of attitudes associated with modern communications to track enforcement officers and systems for jettisoning or concealing irregularities with gear and fish sizes, etc. means that the chances of detection are low and successful prosecution even lower. I do not believe that any CFP regulations are effectively enforced in any sea area, and while allegations of flagrant violation achieved a high profile in the EU/Canada dispute, the incentive to disregard regulations is common to fishermen everywhere.

Being the people they have to be also means that fishermen are likely to be more disposed to take advantage of any opportunities that occur. But they are also victims of the system itself. Collectively they may be stakeholders, but individually they are not. It is every man for himself, and so it is difficult to blame them, even though their attitude is one of the reasons for the ineffectiveness of the conservation regime.

It is difficult to overstate the damage. One has only to look at the direct landings of 'blackfish' or discarding. Fishermen claim they are unable to avoid overquota fish in a mixed fishery, but there is little doubt that by careful selection of gears and seasons based on their own self-avowed skills, they could do more. For example, there can be little justification for confusing saithe with other species after an initial trial shot on promising signs of fish.

The truth is of course that fishermen like to be able to select from a big bag of fish and try to land anything they cannot bring themselves to discard. There is pressure to prohibit discards of overquota and undersized fish, but this would multiply the already difficult and costly task of enforcement. And, if it created any market for undersized fish, then a ban would only add to the pressure to reduce effective mesh sizes. In fact, a solution to that particular problem is ready to hand – fishermen need only increase their mesh sizes, or introduce square mesh nets, to retain only those legal sized fish that make best use of existing market opportunities.

A successful policy depends on enforcement, and successful enforcement depends on the compliance that comes from a sense of ownership and therefore commitment by fishermen. If, as I suggest, *enforced* compliance is impractical, or involves unacceptable cost, then the *only* alternative is voluntary compliance and that can *only* be achieved by a level of allocation of resource ownership that is more meaningful than at present.

Ownership

The maintenance of the principle of relative stability has been a key achievement of the CFP. It does offer some sense of ownership at a national level which has proved of immense value in negotiation of reasonable entitlements in the accession of Spain and Portugal, although it was less helpful between the EU and Norway. Useful though it has been, the sense of ownership embodied in shares based on relative stability is collective and defined only in terms of catching opportunities rather than possession of the resource itself. It does nothing for the individual fishermen of any country and it does nothing to ensure that they can derive long-term benefit from any conservation action they themselves may take. The present system does not provide them with any sense of identity as stakeholders, however much they might wish that it did.

Where ownership exists only in terms of an unallocated catch it leads to naked competition for the best fishing areas at the earliest fishable time in a season, and with the likelihood (certainty) that efficient modern technology will prevail over traditional methods, even though these may be more benign in conservation terms. The development of twin trawl rigs and the cost advantage of gill-netting are examples in EU waters, and the halibut and herring roe fisheries of the north Pacific are more extreme examples of competition within an unallocated ceiling. The competitive edge will always go to the less scrupulous and then competition becomes simply a form of legitimized dispossession. The definition of ownership is therefore critical to enforcement and the control of exploitation.

The point here is that the underlying philosophy of the Treaty of Rome, the Rights of Establishment, and the Single European Act reflect the centralized *communitaire* approach to ownership which may be suitable for countries as an entity, but fishermen who are required to comply need something in addition and at a lower level that they can relate to individually. The concept of free access to fisheries and the accent on competition which goes with it runs counter to the practical necessity of allocating the catch in defence of essential interests. The allocation is, in fact, an attempt to provide fishermen with a notional sense of ownership, but one which means very little because neither they nor the country really have the means to protect that ownership if it comes under pressure. The furore over quota hoppers, legally usurping UK quotas, and 'flags of convenience' in the high seas salmon fishery, are cases in point. One might add that the enhanced mobility of fishermen has exacerbated the ownership problem. Localized fishermen are incensed by what they see as marauding fishermen from elsewhere, even though the intruders may have a quota allocation, and even more so if they have not. The local group's position echoes the older controversy between coastal states and 'distant water' interests, and there may yet be lessons to learn from the solution of extended limits that had to be adopted at that time. There is a contradiction between the centralist aspirations of the EU and the Treaty of Rome, and the localized ownership which is essential to promote an individual sense of responsibility for conservation.

Structure and investment

International discussion of the catch versus effort options for the control of exploitation in NEAFC, but more especially within ICNAF, in the 1960s concluded that since the effectiveness of effort can only be defined in terms of catch, then one might as well regulate catches in the first place. More to the point, catch was the only form of control that one of the major players of the time, the USSR, was prepared to agree to. The first TAC was introduced for George's Bank haddock in 1971, initiating the TAC based system eventually adopted by the EU. It was directed solely at the level of production.

The capital investment in vessels has therefore never been geared to the capacity of the resources to withstand exploitation, but rather to the ability of fishermen to secure loans and the frequently generous credit terms from governments seeking to support their ship-building industries. Loan agencies and ship-builders have never had any regard for the impact of the investment on the resources themselves, and so investment has depended on fishermen being able to persuade the financiers that they can service the loan from the value of catches. The *laissez faire* attitude to the consequences of uncontrolled investment may be close to normal commercial practice in a market economy, but it simply escalates the competition which time and again has been shown to be so damaging to the resources themselves.

The MAGP is a laudable attempt to redress the balance, but its effective implementation is so belated that the benefits will be overtaken by improving technology. Indeed, because of the international fishing capacity calibration problem, the MAGP could never be more than a very blunt instrument and it certainly cannot direct effort away from vulnerable stocks. It may simply increase the net efficiency of the vessels remaining in service. The structural policy itself therefore cannot solve the conservation problem on the timescale required, even though it may promote economic efficiency.

Scientific and management credibility

The weak enforcement referred to above also has an insidious effect on the scientific basis of the regulations themselves, especially the TACs. Holden has discussed resource dynamics and sustainable exploitation. These are fundamental principles which are not negotiable. But as the consequences of disregarding the full intent of scientific evidence become less palatable, so the accuracy of the stock assessments is called in question.

Politicians and industry wishing to favour a generous interpretation of next year's prospects can always dispute scientific estimates by drawing attention to margins of error. This may be correct and acceptable for individual years, but as a persistent year-on-year device for inflating the TACs it goes beyond any justifiable interpretation of the error structures. It becomes yet another way of racking up the level of exploitation, but this time with the tacit endorsement of the managers themselves. That said, the catch figures remain the single most important fact of stock assessment and, even assuming the national statistics of declared catches are accurate, overquota blackfish and discarding have gone.a long way to undermining the credibility of the catch figures. Scientists then have to make allowances for errors in catch recording, which have now become a serious cause for concern and loss of conviction among the scientists themselves.

For years now scientists have sought to overcome errors in the catch figures by cross-checking estimates of stock size against independent resource surveys. But these too are criticized by industry, in particular for not having sufficient regard for the fishermen's knowledge of where the fish are to be found. With modern technology, some fishermen will still be holding that view as they catch the last shoal. Scientists need to know where the fish are, yes, of course, but they also need to know where the fish are *not* compared with some previous year. In fact the individual fisherman's views usually coincide with the scientist's, but this is rather different from an industry's collective perception that needs to have one eye on the political context. Research vessels and their surveys are expensive but absolutely essential underpinning for the whole of the fisheries management programme and need far more support in government funding commitments than they receive at present.

However much the fishermen may appreciate the realities of the situation, they are almost committed to oppose any measure that threatens their own livelihood. Despite the merits of the framework for access and relative stability, the management process is now beset by the industry's perception that, on a year-to-year basis, management is misguided. In their view its only saving grace is that it is ineffectual and managers themselves face a growing loss of conviction in the tools at their disposal. The ease with which both TACs and technical measures can be circumvented has undermined the credibility of the system within countries and between them. As the deterioration of the stocks bites more deeply, so the level of short-term sacrifice required to turn the situation becomes more severe and that, in turn, reinforces the already immensely difficult task of achieving the necessary agreements and implementing them.

The global context

It is not an encouraging outlook, but the EU is not alone. Chronic overfishing is a global problem which has been documented for many years, although the remedy is equally well known. There are deficiencies in the science base, but the underlying problem is the same everywhere – the lack of ownership defined at a level that will provide individual fishermen with a beneficial interest in the well-being of a resource which will encourage compliance with conservation, coupled with the political confidence that regulations could work and have the support of fishermen, rather than persistent confrontation. The only areas of fisheries management that can claim a measure of success are invariably based on relatively sedentary and localized crustacean and shellfish resources which were brought within a regulatory regime at an early stage in their development.

These characteristics are leading to a progressive deterioration on a global scale. It needs only a very weak correlation between the spawning stock size and recruitment to precipitate the calamity of widespread resource collapse which will ring through the ecosystems as a whole, threatening the stability of marine ecosystems on a wider front.

The problems and their pervasive effect have recently been discussed in relation to a particular class of shared resources at the UN Conference on Straddling Stocks and Highly Migratory Species (1995). It is difficult for such a conference to say anything new, but it recognized the necessity to define ownership and provide an effective framework for monitoring and enforcement within existing institutional structures. Laudable as this may be, it harks back to NEAFC and ICNAF which had much the same shortcomings as the CFP.

Options for the future

I believe now that the EU centralized ownership with weak enforcement and only a loosely directed investment scenario precludes effective conservation and therefore the CFP as a whole will inevitably fail, and sooner rather than later. There has to be another way.

One option would place greater emphasis on the direct control of fishing effort. This was originally dismissed in favour of TACs because it can only be established by referring to the catches and because of the difficulties of year-to-year adjustments for technological development. The MAGP is the basis for a move towards a real reduction in fishing, but the need is urgent and on recent experience it is doubtful changes could be agreed on the scale or in the time frame required. Nor would it achieve the fine tuning that will eventually be required to protect preferred high value and therefore sensitive stocks. Holden concludes his book with an exposition of the virtues of a comprehensive EU licensing scheme, and it has many attractions. But while I accept the theory, I do not agree with his conclusions. Direct licensing or any other effort control will suffer a continuing calibration and technological escalation problem and it would still need to be associated with a TAC regime to ensure the correct balance of exploitation between stocks. Furthermore, being a centralized scheme based upon EU ownership of all resources, it would not address the source of the enforcement problem and the cost would therefore be prohibitive. In my view, effort control could be useful as an adjunct to support TACs within the existing regime, but even this would be difficult to achieve, as evidenced by the UK having withdrawn proposals to do just that.

There have also been exhaustive studies of individual transferable quota (ITQ) systems in order to define ownership of the product, at least in more tangible and negotiable detail. ITQs may have the merit of establishing a right to participate in a fishery and, in a very local fishery, this may be synonymous with ownership. But in the more usual EU situation of geographically extended and shared stocks, this will still be a right of participation rather than ownership that establishes the beneficial interest which I regard as the key to cost effective enforcement. Certainly ITQ systems are having some success, but only in countries where the landing points can be easily controlled. The EU is less fortunate: there are innumerable landing points throughout all Member States, and if the broadly allocated TACs cannot be enforced at present, there is no prospect for doing so within an ITQ system. In a multi-national fishery, ITQs would be violated just as frequently as the current TAC, perhaps more so.

There is also no prospect of diverting a significant amount of the present effort to other non-European or deep water species more widely distributed in international waters. The first are already fully exploited and the second would require a redistribution of the existing investment into vessels more suited to the purpose, but against the risks of an uncertain resource base which, for biological reasons, is not thought capable of supporting major fisheries.

Withdrawal from the EU has also been mooted as a solution which would reduce the geographical scale of management responsibility to national waters. This was considered and rejected in the original CFP negotiations, partly because it cut across the emerging philosophy of the EC as it then was, but also because stock migration weakens any concept of national rights based on median lines, etc. It is also argued that the allocation of TACs or effort allocations in such a system would require even more complex multi-bilateral negotiations. This is probably not true because, in practice, the negotiations would have to take place in a forum that could handle all the trade-offs and that would quickly revert to the present system which actually works very well in securing agreements on complex issues; it just breaks down when the agreements have to be enforced.

The difficulties are such that the 'do nothing' option has been resurrected, with the suggestion that the regulatory regime be relaxed, leaving the resources to the whim of market forces. Only the most efficient would survive, leading to some questions on the viability of 'relative stability' and there is also no certainty that the balance between resources and their exploitation would stabilize at a sustainable level. It is widely believed that stocks would collapse or, at best, be held at minimal levels by a small but technologically sophisticated fleet. Most would agree this is not just mischievous speculation, and if it were to happen it would be widely seen as the abrogation of managerial responsibility and a voluntary sacrifice of the marine heritage.

There are no other options on the table at the moment, so there is clearly scope for a new approach, and there may be just such a possibility based on challenging one of the biological over-simplifications underlying the present framework. Stock assessments assume that fish stocks become fully remixed each year and then redistribute themselves along some established biological pattern. This is only partly true for the pelagic species, and it is certainly not true of demersals or molluscs and crustacea. The remixing is a gradual process over a number of years, depending on the mobility of the fish and the level of exploitation itself. Demersal stocks are generally a mosaic of sub-units so that the potential yield from an area is drawn from a set of localized 'sub-stocks' with some measure of migratory exchange with adjacent groups and eventually the whole assemblage that forms a stock. The present approach to pelagic stocks is realistic, but for the demersal, molluscan and crustacea stocks perhaps the biological unit of management and production (i.e. the stock) could be based on partially differentiated sub-stocks which will allow an approach based on a smaller geographical scale.

The management concept that needs to be re-examined under the present circumstances is that the seas are open to fishing unless they are closed. The need for conservation is such that this needs to be reversed so that the fishing grounds should be *closed* unless they are *open*. The open areas can then be seen as 'windows of opportunity'. For the most part, 'open' areas would be open throughout the year, but that need not necessarily be so and some, such as the sandeel fishing areas, might only be opened for a limited period. And it would not be necessary to open the whole of the area occupied by a stock – just a part where fishing could take place. It immediately changes the standpoint of fishing from being an inalienable right to a privileged opportunity in circumstances where it can be permitted under specified conditions.

Going a step further, in basins like the North Sea and the Baltic there must exist a possible balance between a central closed area which could provide a refuge for a core stock of a species, and indeed ecosystem, surrounded by peripheral areas where the balance of migratory exchange within and between years would support localized fisheries. The basic 'closed area' scenario could be based on median lines with internationally contiguous closed areas at the centre of a basin which would provide the refuge and pool of recruits to sustain fisheries in national coastal areas. The areas open to fishing could be bounded by a national fishery limit extended from and along the coastline to whatever distances were agreed as necessary to support a fishery or production target. The size of these closed and open areas would be a source of much debate, but *in principle* the balance must exist. It could take migration rates into account or be selected such that the peripheral fishery(s) could not threaten the integrity of a stock, no matter how hard it is fished on the limited stock in those areas that are open.

The aim would be to place such an 'open' area in the ownership of a suitably constituted authority to fish and manage as it decides, knowing the protection of the stock is ultimately vested in the areas that remain closed. The fishing community of the open areas would then decide their own objectives – to fish on the lines of achieving a sustainable yield for its own fishing area, to sell off opportunities to fishermen from outside the area, or even to fish it out. Given that the relative size of the closed and open areas could be correctly set for the stock(s), then the fishermen could reap the fortunes of their own actions, good or bad, with no grounds for recourse to central government support.

Clearly one should not underestimate the difficulty of setting these areas, but if the size of a closed area could be agreed internationally between Member States, then the detail of the coastal fisheries could be left to the jurisdiction of the country concerned in a way that conforms to the Treaty of Rome as it is expressed within EU principles of subsidiarity (see page 240). It would still be necessary to set annual TACs with Member States responsible for sub-allocation to local fishing interests and sectors, e.g. producer organizations. The scientific assessments would also remain as now in respect of a particular stock, but fishery groups could be encouraged to commission their own independent estimation of the stock in their area as a basis for negotiating their allocation of the national TAC. It is here that the consequences of management within their own group would bite on the individual fishermen.

The problems of enforcement might then also be eased. The area closed to fishing would be closed to fishing by all gears, without derogation, and controlled by satellite surveillance. The criterion of infringement of a closed area would be simply presence or absence in a fishing mode rather than the quantity and composition of any catch which requires real time inspection and the costs that go with it. Access to any local fishery by mobile fishermen from outside the area might be brought into conformity with the Treaty of Rome as a right to negotiate participation, rather than a preserved right to fish. Their entitlement should be a matter for negotiation with the 'owners' and even subject to various controls, e.g. fees. But the burden of local enforcement on both their own members or negotiated participants would fall to the suitably established local fishery group. They should have to ensure their own management regime.

The choice then is between the present, centrally owned 'top down' management which is leading towards a smaller but competitively efficient industry on a European scale, and a 'bottom up' disaggregated set of interdependent but independent fisheries that will do far more to sustain the social fabric of coastal communities in a way that better satisfies other aspects of the Treaty of Rome.

The present system has now fossilized in its own image and, in my view, is bound to fail for the reasons Holden and I have given. My suggestion goes in a different direction from his and will seem far-fetched. But this or something like it is the only alternative to the present concepts. It deserves to be analysed for that reason alone and adapted as necessary to conform with EU legislation and the CFP, perhaps making more extended use of the principle of subsidiarity.

In the early 1900s a low effort, low technology fishery produced 100 000 t of cod from the North Sea each year. The size of the fish is shown in Fig. 14.1. The TAC for 1996 is 50 000 t from a high effort, high technology industry catching the last few cod in the northern North Sea. The Lowestoft angler's fishery no longer exists; the fish do not come; they are not there. The EU fishery resources must not be destroyed simply because the requirements for effective husbandry conflict with the present narrow view of the Treaty of Rome. That would be too absurd. It is not that the Treaty needs changing, it just needs a more creative interpretation that is relevant to the requirements of today rather than yesterday.

Fig. 14.1   The winner! Lowestoft South Pier Angling Competition 1910.

Annex I

Definitions of Technical Terms and Explanations of 'Euro-jargon'

Acquis communitaire: all the legislation adopted by the Community, which a country acceding to the Community has to accept.

Basic regulations: the regulations which lay down the main provisions of a policy in their general detail.

Derogation: a temporary exemption from the need to respect Community legislation.

Exploitation pattern: the rate of fishing mortality on each age group in a stock.

Fishing effort: any measure of fishing activity such as, for example, number of hours fishing, number of hooks fished, kilometres of gillnets; it is broadly but not directly related to fishing mortality rate.

Fishing mortality rate: the proportion of fish dying each year as a result of fishing; in scientific texts it is always denoted as 'F' and expressed as the negative of the power of the exponential function 'e', which gives, confusingly, the number of fish surviving from one year to the next: for example, if $F = 0.2$, the proportion of fish surviving from one year to the next equals $e^{-0.2} = 0.82$ and the proportion of fish dying is $(1 - 0.82) = 0.18$, or 18%. Exponentials are used because they have many mathematical advantages in calculating; in particular they can be added, whereas percentages cannot; in this book all fishing mortality rates are given in annual percentage rates because these more explicitly describe rates of fishing mortality.

Fish stock: in scientific terms a population of a species of fish which is isolated from other stocks of the same species and does not interbreed with them and can therefore be managed independently of other stocks; in Community legislation the term 'stock' is used to mean a species of fish living in a defined sea area (see Annex II), the two not always being synonymous (see also 'management unit').

F_{max}: the fishing mortality rate at which the maximum sustainable yield per recruit will be taken.

$F_{0.1}$: the fishing mortality rate at which the slope of the yield per recruit curve is one tenth of the slope at its origin; in general terms it may be described as the point on the yield per recruit curve at which the yield per recruit is about 90% of the maximum yield per recruit and usually occurs at 50–70% of the fishing mortality rate at which F_{max} occurs.

Luxembourg compromise: the unofficial procedure adopted in 1956 in order to accommodate France under which the Council would not proceed to a vote if any Member State declared that it would be 'against its national interests' to adopt the proposal under discussion; the 'national interests' did not have to be specified: the term derives from the fact that the compromise was agreed at a meeting of the Council in Luxembourg.

Management unit: a species of fish living in a defined sea area (see Annex II) upon which Community management is based (see also 'Fish stock').

Maximum sustainable yield per recruit: the maximum long-term average annual yield per recruit which can be obtained from a stock under any exploitation pattern.

Natural mortality rate: the proportion of fish dying each year from natural causes; it is denoted by 'M' and expressed identically to the fishing mortality rate.

Qualified majority voting: the system of voting in the Community by which most legislation is adopted under Community procedures, 54 votes of the total of 76 votes being needed for adoption.

Recruitment: the process by which young fish enter the fishery, either by becoming large enough to be retained by the gear in use or by migrating from protected areas into areas where fishing occurs.

Recruits: fish which are in the process of becoming vulnerable to being caught (see 'Recruitment').

Spawning stock biomass: the total weight of mature fish in the stock.

Stock biomass: the total weight of all fish in the stock.

Third country: a country which is neither a Member State ('first country'), although never referred to as this, nor a developing country in the Caribbean, African and Pacific areas, mainly former colonial territories of the Member State, with which the Community has a special treaty relationship under the Lomé Convention.

Tonnes, cod equivalent: a factor by which actual tonnages are multiplied in order to convert them into tonnages having approximately equivalent market value, relative to a factor of 1.0 for cod.

Year-class: fish in temperate waters, such as those of the Community, once they are mature, spawn once a year; the progeny resulting from each spawning is called a 'year-class'.

Yield per recruit: the long-term average yield in weight from a stock for every recruit entering the fishery.

Annex II

Acronyms

ACFM: Advisory Committee on Fishery Management of ICES.
IBSFC: International Baltic Sea Fisheries Commission.
ICES: International Council for the Exploration of the Sea.
ICNAF: International Commission for the North-west Atlantic Fisheries.
NAFO: North Atlantic Fisheries Organisation.
NEAFC: North-east Atlantic Fisheries Commission.
NFFO: National Federation of Fishermen's Organisations.
SFF: Scottish Fishermen's Federation.
STCEF: Scientific, Technical and Economic Committee for Fisheries.
STCF: Scientific and Technical Committee for Fisheries.

Annex III

List of Stocks for Which the Community Fixes TACs

The Community fixes TACs for management units, not biological stocks, defined by ICES sub-areas and divisions, plus one division of the Commission for the Central East Atlantic Fisheries (CECAF).

In some cases the management units include two or more stocks; in other cases, one stock is split into two or more management units, usually to meet political requirements, for example, to exclude the vessels of a Member State from fishing in a particular zone.

Stock definitions often include zones in which the species does not occur. This nomenclature was introduced in 1986 in order to prevent fraud because fishermen were reporting species caught from regions not covered by a TAC, in which it was known that the species did not occur, a point upon which it was found difficult to obtain a conviction.

The zones are described in the frontispiece map. The stocks are those listed in Council Regulation No. 3882/91 fixing TACs for 1992.

In the text, the stocks are referred to by the main ICES division in which they occur, for the sake of brevity.

Species	ICES sub-areas and divisions	Species	ICES sub-areas and divisions
Herring	IIIa	Hake	Vb(1), VI, VII, XII, XIV
Herring	IIIb, c, d(1)	Hake	VIIIa, b, d, e
Herring	IIa(1), IVa, b	Hake	VIIIc, IX, X; CECAF
Herring	IVc(2), VIId		34.1.1(1)
Herring	Vb(1), VIa North(3), VIb	Horse mackerels	IIa(1), IV(1)
Herring	VIa South(4), VIIb, c	Horse mackerels	Vb(1), VI, VII, VIIIa, b, d, e,
Herring	VIa Clyde(5)		XII, XIV
Herring	VII, a (6)	Horse mackerels	VIIIc, IX
Herring	VIIe, f	Mackerel	IIa(1), IIIa; IIIb, c, d,(1), IV
Herring	VIIg, h, j, k(7)	Mackerel	II(8), Vb(1), VI, VII, VIIIa, b,
Sprat	IIIa		d, e, XII, XIV
Sprat	IIIb, c, d(1)	Mackerel	VIIIc, IX, X; CECAF
Sprat	IIa(1), IV(1)		34.1.1(1)
Sprat	VIId, e	Plaice	IIIa Skagerrak
Anchovy	VIII	Plaice	IIIa Kattegat
Anchovy	IX, X; CECAF 34.1.1	Plaice	IIIb, c, d(1)
Salmon	IIIb, c, d(1)	Plaice	IIa(1), IV
Capelin	IIb	Plaice	Vb(1), VI, XII, XIV
Cod	IIb	Plaice	VIIa
Cod	IIIa Skagerrak	Plaice	VIIb, c
Cod	IIIa Kattegat	Plaice	VIId, e
Cod	IIIb, c, d(1)	Plaice	VIIf, g
Cod	IIa(1), IV	Plaice	VIIh, k, j
Cod	Vb(1), VI, XII, XIV	Plaice	VIII, IX, X; CECAF
Cod	VIIa		34.1.1(1)
Cod	VIIb, c, d, e, f, g, h, j, k, VIII,	Sole	IIIa; IIIb, c, d(1)
	IX, X; CECAF 34.1.1(1)	Sole	II, IV
Haddock	IIIa; IIIb, c, d(1)	Sole	Vb(1), VI, XII, XIV
Haddock	IIa(1), IV	Sole	VIIa
Haddock	Vb(1), VI, XII, XIV	Sole	VIIb, c
Haddock	VII, VIII, IX, X; CECAF	Sole	VIId
	34.1.1(1)	Sole	VIIe
Saithe	IIa(1), IIIa; IIIb, c, d(1), IV	Sole	VIIf, g
Saithe	Vb(1), VI, XII, XIV	Sole	VIIh, j, k
Saithe	VII, VIII, IX, X; CECAF	Sole	VIIIa, b
	34.1.1(1)	Sole	VIIIc, d, e, IX, X; CECAF
Pollack	Vb(l), VI, XII, XIV		34.1.1(1)
Pollack	VII	Megrim	Vb(1), VI, XII, XIV
Pollack	VIIIa, b	Megrim	VII
Pollack	VIIIc	Megrim	VIIIa, b, d, e
Pollack	VIIId	Megrim	VIIIc, IX, X; CECAF
Pollack	VIIIe		34.1.1(1)
Pollack	IX, X; CECAF 34.1.1(1)	Anglerfish	Vb(1), VI, XII, XIV
Norway pout	IIa(1), IIIa; IV(1)	Anglerfish	VII
Blue whiting	IIa(1), IV(1)	Anglerfish	VIIIa, b, d
Blue whiting	Vb(1), VI, VII	Anglerfish	VIIIe
Blue whiting	VIIIa, b, d	Anglerfish	VIIIc, IX, X; CECAF 34.1.1(1)
Blue whiting	VIIIe	'Penaeus'	
Blue whiting	VIIIc, IX, X; CECAF	shrimps	French Guyana
	34.1.1(1)	Northern deep	
Whiting	IIIa	water prawn	IIIa Skagerrak
Whiting	IIa(1), IV	Norway lobster	IIIa; IIIb, c, d(1)
Whiting	Vb(1), VI, XII, XIV	Norway lobster	IIa(1); IV(1)
Whiting	VIIa	Norway lobster	Vb(1), VI
Whiting	VIIb, c, d, e, f, g, h, j, k	Norway lobster	VII
Whiting	VIII	Norway lobster	VIIIa, b
Whiting	IX, X; CECAF 34.1.1(1)	Norway lobster	VIIIc
Hake	IIIa; IIIb, c, d(1)	Norway lobster	VIIId, e
Hake	IIa(1), IV(1)	Norway lobster	IX, X; CECAF 34.1.1(1)

Notes:

(1) EC zone

(2) Except Blackwater stock: reference is to the herring stock in the maritime region of the Thames Estuary within a zone delimited by a line running due south from Landguard Point (51°56'N, 1°19.1'E) to latitude 51°33'N and thence due west to a point on the coast of the United Kingdom.

(3) Reference is to the herring stock in ICES division VIa, north of 56°00'N and in that part of VIa which is situated east of 07°00'W and north of 55°00'N, excluding the Clyde.

(4) Reference is to the herring stock in ICES division VIa, south of 56°00'N and west of 07°00'W.

(5) Clyde stock: reference is to the stock of herring in the maritime area situated to the north-east of a line drawn between Mull of Kintyre and Corsewall Point.

(6) ICES division VIIa is reduced by the area added to the Celtic Sea bounded:

- to the north by latitude 52°30'N,
- to the south by latitude 52°00'N,
- to the west by the coast of Ireland,
- to the east by the coast of the United Kingdom.

(7) Increased by zone bounded:

- to the north by latitude 52°30'N,
- to the south by latitude 52°00'N,
- to the west by the coast of Ireland,
- to the east by the coast of the United Kingdom.

(8) Excluding EC zone.

References

Only a very limited number of references is provided. In particular, references are given to only the most important pieces of Community legislation.

Books and scientific papers

Anderson, L.G.(1977) *The Economics of Fisheries Management*. The John Hopkins University Press, Baltimore.

Churchill, R.R.(1987) *EEC Fisheries Law*. Martinus Nijhoff Publishers, Dordrecht.

Commission of the European Communities (1981) *Quotas 1981*. SEC (81) 105.

Commission of the European Communities (1991) *Report 1991 from the Commission to the Council and the European Parliament on the Common Fisheries Policy*. SEC (91) 2288 final.

Commission of the European Communities (1992) *Report on monitoring implementation of the Common Fisheries Policy*. SEC (19) 394 final.

Commission of the European Communities (1992) *Report from the Commission to the Council on the discarding of fish in Community fisheries: causes, impact, solutions*. SEC (92) 423 final.

Farnell, J. and Elles, J. (1984) *In Search of a Common Fisheries Policy*. Gower Publishing, Aldershot.

Graham, M. (1935) Modern theory of exploiting a fishery, and application to North Sea trawling. *J. Cons. int. Explor. Mer*, 10, 264–274.

Hannesson, R. (1993) *Bioeconomic Analysis of Fisheries*. Fishing News Books, Oxford.

Holden, M.J. (1978) Long-term changes in landings of fish from the North Sea. In *North Sea fish stocks – recent changes and their causes* (Ed. G. Hempel). *Rapp. P-v. Réun. Cons. int. Explor. Mer*, 172, 11–26.

Jones, R. and Hislop, J.R.G. (1978) Changes in North Sea haddock and whiting. In *North Sea fish stocks – recent changes and their causes* (Ed. G. Hempel). *Rapp. P-v. Réun. Cons. int. Explor. Mer*, 172, 58–71.

Leigh, M. (1983) *European Integration and the Common Fisheries Policy*. Croom Helm, Beckenham.

Lockwood, S.J. (1988) *The Mackerel: its biology, assessment and the management of the fishery*. Fishing News Books, Oxford.

Legislation

Council Regulation (EEC) No. 100/76 of 19 January 1976 on the common organization of the market in fishery products. OJEC, **19**, L20, 1–18, 28.1.1976

Council Regulation (EEC) No. 101/76 of 19 January 1976 laying down a common structural policy for the fishing industry. OJEC, **19**, L20, 19–22, 28.1.1976

Council Regulation (EEC) No. 2057/82 of 29 June 1982 establishing certain control measures for fishing activities by vessels of the Member States. OJEC, **25**, L220, 1–5, 29.7.1982

Council Regulation (EEC) No. 170/83 of 25 January 1983 establishing a Community system for the conservation and management of fishery resources. OJEC, **26**, L24, 1–13, 27.1.1983

Council Regulation(EEC) No. 171/83 of 25 January 1983 laying down certain technical measures for the conservation of fishery resources. OJEC, **26**, L24, 14–29, 27.1.1983

Council Regulation (EEC) No. 172/83 of 25 January 1983 fixing for certain fish stocks and groups of fish stocks occurring in the Community's fishing zone, total allowable catches for 1982, the share of these catches available to the Community, the allocation of that share between the Member States and the conditions under which the total allowable catches may be fished. OJEC, **26**, L24, 30–67, 27.1.1983

Council Regulation (EEC) No. 3094/86 of 7 October 1986 laying down certain technical measures for the conservation of fishery resources. OJEC, **29**, L288, 1–20, 11.10.1986

Council Regulation (EEC) No. 4028/86 of 18 December 1986 on Community measures to improve and adapt structures in the fisheries and aquaculture sector. OJEC, **29**, L376, 7–24, 31.12.1986

Council Regulation (EEC) No. 2241/87 of 23 July 1987 establishing certain control measures for fishing activities. OJEC, **30**, L207, 1–7, 29.7.1987

Council Regulation (EEC) No. 345/92 of 27 January 1992 amending for the eleventh time Regulation No. 3094/86 laying down certain technical measures for the conservation of fishery resources. OJEC, **35**, L42, 1–23, 18.2.1992

Council Regulation (EEC) No. 3759/92 of 17 December 1992 on the common organization of the market in fishery and aquaculture products. OJEC, **35**, L388, 1–36, 31.12.1992

Council Regulation (EEC) No. 3760/92 of 20 December 1992 establishing a Community system for fisheries and aquaculture. OJEC, **35**, L389, 1–14, 31.12.1992

Index